Topics in Current Physics

33

Topics in Current Physics   Founded by Helmut K. V. Lotsch

1 **Beam-Foil Spectroscopy**
Editor: S. Bashkin

2 **Modern Three-Hadron Physics**
Editor: A. W. Thomas

3 **Dynamics of Solids and Liquids by Neutron Scattering**
Editors: S. W. Lovesey and T. Springer

4 **Electron Spectroscopy for Surface Analysis**
Editor: H. Ibach

5 **Structure and Collisions of Ions and Atoms**
Editor: I. A. Sellin

6 **Neutron Diffraction**
Editor: H. Dachs

7 **Monte Carlo Methods** in Statistical Physics
Editor: K. Binder

8 **Ocean Acoustics**
Editor: J. A. DeSanto

9 **Inverse Source Problems** in Optics
Editor: H. P. Baltes

10 **Synchrotron Radiation**
Techniques and Applications
Editor: C. Kunz

11 **Raman Spectroscopy** of Gases and Liquids
Editor: A. Weber

12 **Positrons in Solids**
Editor: P. Hautojärvi

13 **Computer Processing of Electron Microscope Images**
Editor: P. W. Hawkes

14 **Excitons** Editor: K. Cho

15 Physics of **Superionic Conductors**
Editor: M. B. Salamon

16 **Aerosol Microphysics I**
Particle Interactions
Editor: W. H. Marlow

17 **Solitons**
Editors: R. Bullough, P. Caudrey

18 **Magnetic Electron Lenses**
Editor: P. W. Hawkes

19 Theory of **Chemisorption**
Editor: J. R. Smith

20 **Inverse Scattering Problems** in Optics
Editor: H. P. Baltes

21 **Coherent Nonlinear Optics**
Recent Advances
Editors: M. S. Feld and V. S. Letokhov

22 **Electromagnetic Theory of Gratings**
Editor: R. Petit

23 **Structural Phase Transitions I**
Editors: K. A. Müller and H. Thomas

24 **Amorphous Solids**
Low-Temperature Properties
Editor: W. A. Phillips

25 **Mössbauer Spectroscopy II**
The Exotic Side of the Method
Editor: U. Gonser

26 **Crystal Cohesion and Conformational Energies** Editor: R. M. Metzger

27 **Dissipative Systems in Quantum Optics** Resonance Fluorescence
Optical Bistability, Superfluorescence
Editor: R. Bonifacio

28 **The Stratospheric Aerosol Layer**
Editor: R. C. Whitten

29 **Aerosol Microphysics II**
Chemical Physics of Microparticles
Editor: W. H. Marlow

30 **Real-Space Renormalization**
Editors: T. W. Burkhardt and J. M. J. van Leeuwen

31 **Hyperfine Interactions of Radioactive Nuclei**
Editor: J. Christiansen

32 **Superconductivity in Ternary Compounds I**
Structural, Electronic, and Lattice Properties
Editors: Ø. Fischer and M. B. Maple

33 **Molecular Collision Dynamics**
Editor: J. M. Bowman

34 **Superconductivity in Ternary Compounds II**
Superconductivity and Magnetism
Editors: M. B. Maple and Ø. Fischer

Molecular Collision Dynamics

Edited by J.M. Bowman

With Contributions by
M. Baer J.M. Bowman G.C. Schatz
R. Schinke D. Secrest

With 38 Figures

Springer-Verlag Berlin Heidelberg New York 1983

Professor J. M. Bowman
Department of Chemistry, Illinois Institute of Technology, Chicago, IL 60616, USA

ISBN-13:978-3-642-81945-2 e-ISBN-13:978-3-642-81943-8
DOI: 10.1007/978-3-642-81943-8

2153/3130-543210

Preface

This monograph covers a broad spectrum of topics in the very broad field of gas phase molecular collision dynamics. The Introduction previews each of the four following topics and attempts to sew them together with a common thread. In addition, a brief review of quantum reactive scattering is given there along with some general remarks which highlight the difficulties in doing quantum reactive scattering calculations.

The chapters are all written by theoreticians who are, of course, experts in the subjects they have written about. Three chapters, the ones by Secrest, Schatz, and the one by Schinke and Bowman deal with non-reactive atom-molecule scattering. Collectively, they describe nearly the full breadth of scattering methods in use today, from fully quantum mechanical to semiclassical and quasiclassical.

The chapter by Baer is the only one dealing with quantum reactive scattering with the additional complexity of the coupling of two potential energy surfaces. The one simplifying feature of the treatment is that the reaction is constrained to be collinear.

Overall, this monograph is mainly a review of the recent advances in the field of molecular collision dynamics, with, however, a considerable amount of new material.

It is hoped that workers and students in the field will find reading the monograph both enlightening and enjoyable.

Chicago, October 1982 *J.M. Bowman*

Contents

List of Contributors

Baer, Michael
Soreq Nuclear Research Centre, Theoretical Chemistry Unit, Yavne, and Department of Chemical Physics, The Weizmann Institute for Science, Rehovot,Israel

Bowman, Joel M.
Department of Chemistry, Illinois Institute of Technology, Chicago, IL 60616, USA

Schatz, George C.
Department of Chemistry, Northwestern University, Evanston, IL 60201, USA

Schinke, Reinhard
Max-Planck-Institut für Strömungsforschung, Böttingerstraße 4 - 8, D-3400 Göttingen, Fed. Rep. of Germany

Secrest, Don
School of Chemical Sciences, Urbana, IL 61801, USA

1. Introduction

J.M. Bowman

This book contains four chapters devoted to important topics in molecular collision dynamics. They deal with relatively new topics or with an older one but in a new way. They are all strictly theoretical, however, in every case it is easy to point to experiments which are directly related to the theoretical discussions. Chapter 3 especially contains numerous comparisons with and references to experiment.

The theoretical description of electronically adiabatic nonreactive atom-molecule and molecule-molecule scattering is well understood. Unfortunately, the general numerical implementation of the theory is hindered by the present speed and size of computers. Also unfortunate, but in no way the fault of computers, is the vast amount of information contained in the exact scattering wave function compared to what is measured experimentally (perhaps, this is the fault of the experimentalists!). Thus, by necessity and perhaps even by preference, one is forced into the rich area of approximate theories.

It is this area which is the subject of this book. In each chapter a different approximation method is discussed and applied to a different class of collision systems.

In Chap.2, D. Secrest reviews several reduced dimensionality quantum approximations, focusing mainly on the most accurate one, the coupled states approximation which was developed by MCGUIRE and KOURI [1.1], KOURI and co-workers [1.2] and PACK [1.3]. He then goes on to present some important new material on perturbation corrections to the coupled states approximation. I found this discussion of the coupled states approximation to be particularly lucid.

The reduced dimensionality quantum theories of nonreactive atom-diatom scattering will eventually be applied to vibrational and rotational energy transfer in polyatomic systems. However, at present the large number of coupled states still presents a formidable computational barrier. For such systems the quasiclassical trajectory approximation is expected to be well-suited and seemingly straightforward to apply. This subject is discussed by G.C. Schatz in Chap.3. Schatz begins by reviewing the previous experimental and theoretical literature in this area and then focusses on his own quasiclassical studies. He addresses the important (and previously glossed-over) issues of the efficient determination of good semiclassical vibrational states and the treatment of vibration/rotation coupling in collision systems. Numerical results are presented for He+SO_2 and Kr+CO_2. There is also a very useful discussion and test of a time-saving trajectory sudden approximation. Schatz also provides useful "blueprints" to aid the serious workers in the field who may wish to begin their own studies of energy transfer in polyatomic systems.

As I mentioned, we believe that the quasiclassical trajectory method is well-suited to study collision systems where many states are coupled together. To be a bit more precise, I should add the condition that the objects of interest, e.g., cross sections, rate constants, involve summing over many final states and averaging over many initial ones. As a consequence of this extensive summing and averaging, it had been argued that interesting coherent quantum effects would be substantially quenched out and that the (incoherent) quasiclassical description would be quite adequate. This argument had been made even for atom-rigid rotor systems. Very recently, however, and in my opinion, happily, experimental and theoretical evidence of distinct and sharp features in rotationally inelastic differential cross sections have appeared. These features, termed rotational rainbows, are discussed in Chap.4 by R. Schinke and J.M. Bowman. Both a review of the theory of these features as well as an extensive discussion of the experiments is given. The theory is based on one of the reduced dimensionality quantum approximations discussed by Secrest in Chap.2, the infinite-order sudden approximation. This is an approximation with a rich history and many names associated with it. In the 1979 review by KOURI [1.2], forty-five references were cited in connection with the origins and "latest" modifications of this theory (the list is certainly much larger now!). However, the works of PACK [1.3] and SECREST [1.4] are generally regarded as the central ones in the time-independent theory. Independently, SCHINKE [1.5] and BOWMAN [1.6] investigated the classical limit of this theory and, using it, elucidated the physics of rotational rainbows. THOMAS [1.7], in an earlier trajectory study of Li^+ + CO, argued for their existence and must be credited for coining the term.

Chapter 5 by M. Baer stands somewhat alone in this book, dealing with electronically nonadiabatic quantum reactive scattering. This reminds us that there are important problems in molecular collision dynamics which are several levels of complexity beyond what I have been discussing thus far. Baer has bravely (and almost single-handedly) worked in this field and presents a very nice review of the theory of these processes as well as several interesting numerical studies (albeit confined to the collinear world).

Quantum reactive scattering remains a stubborn problem, although progress in this area is being mode. The rest of this Introduction will review the recent progress in this field.

A number of complementary review articles on quantum reactive scattering have appeared recently. In a 1979 review, CONNOR [1.8] surveyed nearly every aspect of reactive scattering including the generation and fitting of *ab initio* and semiempirical potential energy surfaces and references to the experimental literature. It is perhaps the most sweeping review of the field available. A more recent comprehensive review of reactive scattering, with more emphasis on three-dimensional quantum reactive scattering, has been given by WALKER and LIGHT [1.9]. A number of valuable, more specialized reviews of quantum reactive scattering are also available. WYATT's [1.10,11] 1979 reviews examine accurate and approximate three-dimensional calculations of the $H + H_2$ reaction in detail [1.10] as well as numerous approximate quantum reactive scattering theories with emphasis on several reactive systems [1.11]. A detailed review and discussion by KUPPERMANN [1.12] of the accurate quantum reactive scattering methods and calculations of the Kuppermann group through 1981 is particularly suitable for newcomers to the field. The didactic review by LIGHT [1.13] is also recommended for nonspecialists. The most recent overview of the field can be found in a review by SCHATZ [1.14]. A comparison of the latest theoretical and experimental rate constants for the $H + H_2$ and $D + H_2$ reactions which is given there, points to a discrepancy between theory and experiment for the reaction with vibrationally excited H_2. Thus, even for the "simplest" chemical reaction, some unresolved questions remain.

There have been several recent advances in the field. LIGHT and WALKER [1.15] made a considerable advance in the efficiency of reactive scattering calculations by propagating the R-matrix instead of the scattering wave function. The efficiency of their method, which is inherently stable, is due mainly to the elimination of the time-consuming stabilization transformations which are required by wave function propagation methods. The instability of these methods arises because of the need to propagate energetically closed channels which are added to the basis representation of the wave function to ensure completeness. It is this completeness requirement which is the most difficult one to fulfill in reactive scattering. Until very recently, the problem was tackled by expressing the Hamiltonian in terms of reaction coordinates. In practice, the configuration space is subdivided and described by cartesian coordinates (for reactants) in one subregion, polar coordinates in other regions and finally cartesian coordinates again (for the products). The switching of coordinates systems as well as the use of polar coordinates are at once an answer to the completeness problem and the source of much of the attendant computational labor. Very recently, KUPPERMANN et al. [1.16] and HAUKE et al. [1.17] have employed a single, hyperspherical coordinate system, introduced by DELVES [1.18,19], in studies of collinear reactions. As noted by these authors, this system of coordinates has several advantages over the previous piecewise ones. First, the entire reactive scattering space, including dissociation, can be spanned by a denumerable set of basis functions. Second, the basis functions obtained in the hyperspherical coordinate system is apparently more flexible than the previous ones based on reaction coordinates. This accounts for the relative ease with which small skew angle systems, e.g., I + HI [1.20,21], can be dealt with in the former system as well as the more rapid convergence of the scattering matrix with respect to basis size.

The technology to describe quantum reactive scattering in two mathematical dimensions, e.g., collinear reactions, is fairly well advanced. Such calculations require modest amounts of computer time relative to full three-dimensional calculations (where the technology is still under development). Thus, it seems natural to develop approximate three-dimensional quantum theories of reactive scattering which incorporate reduced dimensionality, i.e., collinear, quantum calculations. Perhaps the first attempt to do this was made by BERNSTEIN and LEVINE [1.22]. The conversion of collinear to three-dimensional reaction probabilities is accomplished essentially by a rescaling of the collinear world phase space volume by the three-dimensional one and assuming an invariance of the collinear surprisal. The transformation is justified by involing general information-theoretic arguments which have been used to describe many collision processes (for a recent review of this work, see [1.23] by LEVINE and KINSEY). BAER [1.24] has considered somewhat related ideas; both his approach and the one of LEVINE and BERNSTEIN consider the effects of noncollinear degrees-of-freedom of either the separated reactants or products on the collinear transition probabilities.

We have taken a different point of view in developing a theory which incorporates reduced dimensionality exact quantum reaction probabilities into three-dimensional transition state theory [1.25]. Both integral and differential cross sections are obtained which are vibrationally "state-to-state" but implicitly energy-averaged over initial rotational states and summed over final ones. The theory has been tested very successfully against the three-dimensional accurate quantum cross sections of SCHATZ and KUPPERMANN [1.26] for the $H + H_2(\nu = 0)$ reaction [1.27]. Differential cross sections for this reaction are also in good agreement with the accurate ones. The expression for the differential cross section involves the collinear scattering matrices which for many reactions exhibit marked resonance features. We have very recently used the collinear exact quantum scattering matrices of KAYE and KUPPERMANN [1.28] for the $F + H_2$, $F + D_2$, and $F + HD$ reactions to obtain differential cross sections for formation of various final vibrational states of the products [1.29,30]. For those reactions with a marked collinear resonance, the corresponding differential cross sections exhibit sharp, distinguishing features relative to those cross sections for reactions without collinear resonances. The presence of these features and their qualitative detail are confirmed by the recent experiments by the LEE group [1.31,32]. Very important additional and supporting theoretical evidence of resonant behavior in the three-dimensional $F + H_2$ reaction can be found in the pioneering work of REDMON and WYATT [1.33] and in the recent infinite-order sudden calculations of JELLINEK et al. [1.34].

Thus, progress in the field has been made and with continued progress in both theory and detailed molecular beam experiments, the future looks bright indeed for a clear and intimate view of molecular collision dynamics.

References

1.1 P. McGuire, D.J. Kouri: J. Chem. Phys. *60*, 2488-2499 (1974)
1.2 D.J. Kouri: "Rotational Excitation II: Approximation Methods", in *Atom-Molecule Collision Theory*, ed. by R.B. Bernstein (Plenum, New York, London 1979) Chap.9, pp.301-358
1.3 R.T. Pack: J. Chem. Phys. *60*, 633-639 (1974)
1.4 D. Secrest: J. Chem. Phys. *62*, 710-719 (1975)
1.5 R. Schinke: Chem. Phys. *34*, 65-79 (1978)
1.6 J.M. Bowman: Chem. Phys. Lett. *62*, 309-311 (1979)
1.7 L.D. Thomas: J. Chem. Phys. *67*, 5224-5236 (1977)
1.8 J.N.L. Connor: Comp. Phys. Comm. *17*, 117-144 (1979)
1.9 R.B. Walker, J.C. Light: Ann. Rev. Phys. Chem. *31*, 401-433 (1980)
1.10 R.E. Wyatt: "Direct-Mode Chemical Reactions I: Methodology for Accurate Quantal Calculations", in *Atom-Molecule Collision Theory*, ed. by R.B. Bernstein (Plenum, New York, London 1979) Chap.17, pp.567-594
1.11 R.E. Wyatt: "Reactive Scattering Cross Sections II: Approximate Quantal Treatments", in *Atom-Molecule Collision Theory*, ed. by R.B. Bernstein (Plenum, New York, London 1979) Chap.15, pp.477-503
1.12 A. Kuppermann: Theor. Chem. *6*A, 79-164 (1981)
1.13 J.C. Light: "Reactive Scattering Cross Sections I: General Quantal Theory", in *Atom-Molecule Collision Theory*, ed. by R.B. Bernstein (Plenum, New York, London 1979) Chap.14, pp.467-476
1.14 G.C. Schatz: "Overview of Reactive Scattering", in *Potential Energy Surfaces and Dynamics Calculations*, ed. by D.G. Truhlar (Plenum, New York, London 1981) Chap.12, pp.287-310
1.15 J.C. Light, R.B. Walker: J. Chem. Phys. *65*, 4272-4282 (1976)
1.16 A. Kuppermann, J.A. Kaye, J.P. Dwyer: Chem. Phys. Lett. *74*, 257-262 (1980)
1.17 G. Hauke, J. Manz, J. Römelt: J. Chem. Phys. *73*, 5040-5044 (1980)
1.18 L.M. Delves: Nucl. Phys. *9*, 391-399 (1959)
1.19 L.M. Delves: Nucl. Phys. *20*, 275-308 (1960)
1.20 J.A. Kaye, A. Kuppermann: Chem. Phys. Lett. *77*, 573-579 (1981)
1.21 J. Manz, J. Römelt: Chem. Phys. Lett. *81*, 179-184 (1982)
1.22 R.B. Bernstein, R.D. Levine: Chem. Phys. Lett. *29*, 314-318 (1974)
1.23 R.D. Levine, J.L. Kinsey: "Information-Theoretic Approach to Molecular Collisions", in *Atom-Molecule Collision Theory*, ed. by R.B. Bernstein (Plenum, New York, London 1979) Chap.22, pp.693-750
1.24 M. Baer: J. Chem. Phys. *62*, 4545-4550 (1975)
1.25 J.M. Bowman, J.-G. Zhi, K.-T. Lee: J. Phys. Chem. *86*, 2232-2239 (1982)
1.26 G.C. Schatz, A. Kuppermann: J. Chem. Phys. *65*, 4668-4692 (1976)
1.27 J.M. Bowman, G.-Z. Ju, K.-T. Lee: J. Chem. Phys. *75*, 5199-5201 (1981)
1.28 A. Kuppermann, J.A. Kaye: J. Phys. Chem. *85*, 1969-1972 (1981)
1.29 J.M. Bowman, K.-T. Lee, J.-G. Zhi: Chem. Phys. Lett. *86*, 384-388 (1982)
1.30 K.-T. Lee, J.M. Bowman: J. Phys. Chem. *86*, 2289-2291 (1982)
1.31 R.K. Sparks, C.C. Hayden, K. Shobatake, D.M. Neumark, Y.T. Lee: "Molecular Beam Studies of Reaction Dynamics of F + H_2, D_2", in *Horizons of Quantum Chemistry*, ed. by K. Fukui, B. Pullman (Reidel, Dordrecht 1980) pp.91-105
1.32 Y.T. Lee: Private communication
1.33 M.J. Redmon, R.E. Wyatt: Chem. Phys. Lett. *63*, 209-212 (1979)
1.34 J. Jellinek, M. Baer, D.J. Kouri: Phys. Rev. Lett. *47*, 1588-1592 (1981)

2. Inelastic Vibrational and Rotational Quantum Collisions*

D. Secrest

Scattering theory and related scattering measurements have made dramatic strides in the last few years. There are a number of excellent reviews in various areas of this field [2.1-5]. The entire field of atom-molecule scattering theory has been rather thoroughly treated in the excellent book *Atom-Molecule Collision Theory*, edited by BERNSTEIN [2.6]. This book covers many areas, from the computation of interaction potentials through numerical techniques for both classical and quantum mechanical calculations and approximation methods. Since another review is not needed at this time, I intend to concentrate here on ideas and methods which I believe will lead to an understanding of the details of energy transfer in collisions and eventually to understanding chemical reactions. I do not intend to cover all or even a majority of the considerations which will contribute to such an understanding. On the contrary, I shall mainly concern myself with the accurate calculation of rotational and vibrational energy transfer in atom-molecule collisions. Accurate calculation does not necessarily imply the solution of the close coupled equations. In fact, if we limit ourselves to solution of the close coupled equations, we restrict the problems which we may study to a pitful few. It is, of course, important that this select group of problems be solved accurately as benchmarks for future work and to serve as a source of experience on which to develop intuition. The central problem, however, is to devise approximations which will allow us to handle larger systems accurately. A system of approximations is needed which will allow us to describe each feature of a collision in which we have an interest in the desired accuracy. There are a few approximations, developed in recent years, which represent a step in the right direction. The most fruitful approach at present appears to be the dimensionality reducing approximations pioneered by RABITZ [2.7]. The most popular approximation to result from this approach is the coupled states (CS) approximation [2.8], also referred to as the centrifugal sudden approximation [2.9] or the j_z approximation. This approach is being actively investigated by a number of workers and will be discussed in some detail here. There are many situations in which this approximation does not work, and we are slowly gaining an understanding of the conditions under which the method should be expected

*This work was supported by a grant from the National Science Foundation

to be applicable. The infinite order sudden (IOS) [2.10] and energy sudden (ES) [2.11,12] approximations also reduce the computational effort to the point where very complex systems may be treated. There are many questions about the accuracy of these approximations.

2.1 General Computational Techniques

The coupled scattering equations both for accurate close coupled calculations and many of the approximations have the same form. There has been much work in the last ten years on the solution of these systems of equations with scattering boundary conditions. There were recently two NRCC work shops in which many of the current numerical techniques were discussed and compared. The details of these methods are given in the reports of the workshops [2.13] as well as results of the tests [2.14]. A shortened account of the tests has been published in the open literature [2.15].

No one method was found to be best for all test problems. The methods in current use for the most part fall into four general categories [2.16]. There are two general approaches: the approximate solution approach and the approximate potential approach. The approximate solution approaches are classical numerical solution techniques for solving the differential or corresponding integral equation. The approximate potential approaches consist of approximating the potential over discrete ranges by a simple function for which the solution can be found analytically. For each of these two approaches there are two basic techniques, the solution following technique and the invariant embedding technique. Again, the solution following technique consists of starting the solution usually in the neighborhood of the origin and integrating into the asymptotic region, as in the classical numerical techniques. The violent growth of some channels in the nonclassical region necessitates carrying a complete set of solutions for purposes of stabilizing the solution, as well as for constructing the desired boundary conditions when the asymptotic region is reached. The invariant embedding techniques consist of solving a set of scattering problems which have the same potential as the base problem only over a finite range and zero everywhere else. These elementary solutions are then assembled exactly to give a solution to the base problems [2.16] and [Ref.2.6, p.265]. The invariant embedding technique also develops a complete set of solutions but requires no stabilization. Both the solution following and invariant embedding techniques were found to require about the same amount of computer time in the NRCC tests, but the invariant embedding techniques were found to be somewhat easier to use as it was not necessary to choose stabilization points. The approximate solution approach was found to be most efficient when the potential was rapidly varying, as is usually the case in the strong interaction region of the collision, while the approximate potential approach was the method of choice for slowly varying potentials such as those found in the long-range tail of the interaction. The conclusion of the

NRCC trials was that a hybrid method was best, employing an approximate solution method in the rapidly varying region of the potential and an approximate potential method in the slowly varying region. Of course, this was not earthshaking news as a number of us were aware of this and had been using hybrid methods well before the NRCC workshop. A group of participants in the workshop put together a hybrid program [2.17] called VIVAS which uses the Log-derivative [2.18] method (an approximate solution approach) for the nonclassical region and the variable interval variable step method [2.19] (an approximate potential approach) for the long-range tail of the potential. Both of these methods are invariant embedding techniques. VIVAS appears to be a good general program as a first choice for attempting any new calculation. It may not be the best choice for any particular problem, however. It was not the best even for all of the small set of test problems used in the NRCC tests. For one of the test problems an approximate solution approach in the solution following technique won. The approximate potential approaches are especially efficient when the problem must be solved for more than one energy, as much of the work involved in solving for the first energy may be saved and used for further energies. Thus VIVAS would have beaten the approximate solution approach on the second energy, but the particular test problem used an energy-dependent potential and even VIVAS would in this case need to be restarted for the second energy [2.15].

There are as yet no clear rules governing the most appropriate method for solving a scattering problem numerically. When faced with a new scattering problem, VIVAS is probably a good program to start with. The NRCC study is at present the most definitive study of available methods, but I am sure that improved methods will appear in the years to come. Since the recent reviews [Ref.2.6, p.265] and the NRCC report [2.15] cover the methods for solving the coupled systems adequately, I will limit my discussion of numerical techniques to those not discussed adequately in the recent literature and to recent improvements [2.20] of the CS approximation.

2.1.1 The Scattering Equations

To define our notation, we will start with the close coupling equations for scattering in reduced units in the total angular momentum representation

$$\left[-\frac{\partial^2}{\partial R^2}+\frac{\ell(\ell+1)}{R^2}-k_i^2\right]f_{iI}^J(R)=-\sum_{i'}V_{ii'}^J(R)\,f_{i'I}^J(R) \tag{2.1}$$

with boundary conditions,

$$f_{iI}^J(R)\xrightarrow[R\to 0]{}0$$

and

$$f^J_{iI}(R) \underset{R\to\infty}{\sim} \frac{1}{k_i^{\frac{1}{2}}} \left(\exp[-i(k_iR - \ell\pi/2)]\delta_{iI} - S^J_{iI}\exp[i(k_iR - \ell\pi/2)]\right) . \qquad (2.2)$$

Here i and I are composite subscripts of the final and initial quantum numbers, respectively. For an atom-diatomic molecule collision, for example, $i = (\ell,j,\nu)$, where ℓ is the relative angular momentum of the atom and molecule, j is the angular momentum of the diatomic and ν is the vibrational quantum number which is absent if a rigid rotor is being considered. If the molecule is a polyatomic molecule, there will be of course a k quantum number related to the symmetry state of the polyatomic molecule and a vibrational quantum number for each normal mode considered in the collision. For the collision of two molecules, an angular momentum for each molecule would appear as well as some rather arbitrary intermediate coupling angular momenta. Since the phenomena I shall discuss can be adequately described by an atom-molecule collision, I will restrict the discussion to such systems. The $V^J_{ii'}(R)$ in (2.1) is given by

$$V^J_{ii'}(R) = \int \psi^{JM*}_{\ell j\nu}(\vec{\eta}\alpha\beta\gamma\theta\varphi) qp(\alpha\beta\gamma\theta\varphi R\vec{\eta})\psi^{JM}_{\ell' j'\nu'} d\Omega_R \quad , \qquad (2.3)$$

where

$$\psi^{JM}_{\ell j\nu}(\vec{\eta}\alpha\beta\gamma\theta\varphi) = \sum_{m_j m} (jm_j\ell m|JM) Y^m_\ell(\theta\varphi)\varphi^{jm_j}_\nu(\vec{\eta}\alpha\beta\gamma) \qquad (2.4)$$

and

$$\varphi^{jm_j}_\nu(\vec{\eta}\alpha\beta\gamma) = \sum_k a^{jm_j}_{\nu k} D^j_{m_j k}(\alpha\beta\gamma) H_{jk\nu}(\vec{\eta}) \quad . \qquad (2.5)$$

The $\varphi^{jm_j}_\nu$ is a symmetry function of the molecule. The $\vec{\eta}$ is a set of internal vibrational coordinates of the molecule, the number of components depending on the number of vibrational degrees of freedom of the molecule. The notation $d\Omega_R$ indicates integration over all coordinates except R. The ν quantum number labels the vibrational state. It is often taken to be a composite quantum number with a label for each normal mode of the molecule, but in general it is not possible to separate normal modes of the molecule, except approximately, for the lower vibrational states. The φ^{jm_j} are eigenfunctions of the molecule Hamiltonian which are, in general, chosen so that they transform according to irreducible representations of the symmetry group for the molecule. Since the interaction potential must be invariant to these symmetry operations, the potential matrix will become block diagonal, greatly simplifying the problem. Thus, the $a^{jm_j}_{\nu k}$ are chosen such that φ_ν transforms as a representation of the symmetry group of the molecule and satisfies the Schrödinger equation for the free molecule. This is not a trivial problem in itself. If the molecule is a diatomic molecule, γ may be taken as zero and the sum and k quantum number disappear. It has recently been shown that even for linear vibrating triatomic molecules, the wave function can be exactly represented as in

(2.5) as a product of D matrices with a function of the vibrational modes [2.21]. In this representation, the bending mode is treated as a single mode rather than two degenerate bending modes. The $(jm_j \ell m|JM)$ is a Clebsch-Gordon coefficient.

Using (2.4), the interaction potential may be simplified to some extent. Since the interaction potential is invariant to a coordinate rotation, we may rotate the coordinate system such that θ' and φ' are zero to give

$$V^J_{ii'}(R) = \sum_{m_j m m'_j m'} (jm_j \ell m|JM)(j'm'_j \ell' m'|JM)$$
$$\sum_{n n_j n' n'_j} \int D^{(\ell)^*}_{mn}(\varphi\theta\alpha) D^{(j)^*}_{m_j n_j}(\varphi\theta\alpha) D^{(\ell')}_{m'n'}(\varphi\theta\alpha) D^{(j')}_{m'_j n'_j}(\varphi\theta\alpha) Y^{n^*}_{\ell}(00) Y^{n'}_{\ell'}(00)$$
$$\left[\int \varphi^{jn_j^*}_{\nu}(\vec{\eta}0\beta'\gamma') V(0\beta'\gamma' 00R\vec{\eta}) \varphi^{j'n'_j}_{\nu'}(\vec{\eta}0\beta'\gamma') d\Omega_R\right] \quad (2.6)$$
$$\sin\theta \, d\theta \, d\varphi \, d\alpha \quad .$$

The definition of the rotation matrices used here is that of EDMONDS [2.22] but the final results will be the same with any definition of the rotation matrices.

$$V^J_{ii'} = \sum_{m_j m m'_j m' n_j n'_j} (jm_j \ell m|JM)(j'm'_j \ell' m'|JM)$$
$$\int D^{(\ell)^*}_{m0}(\varphi\theta\alpha) D^{(\ell')}_{m'0}(\varphi\theta\alpha) D^{(j)^*}_{m_j n_j}(\varphi\theta\alpha) D^{(j')}_{m'_j n'_j}(\varphi\theta\alpha) \frac{[(2\ell+1)(2\ell'+1)]^{\frac{1}{2}}}{4\pi} \sin\theta \, d\theta \, d\varphi \, d\alpha$$
$$\left(\frac{1}{2\pi}\right) V^{n_j n'_j}_{\nu j \nu' j'} \quad . \quad (2.7)$$

The integral in (2.7) can be performed in two different ways to give

$$\int D^{(\ell)^*}_{m0}(\varphi\theta\alpha) D^{(\ell')}_{m'0}(\varphi\theta\alpha) D^{(j)^*}_{m_j n_j}(\varphi\theta\alpha) D^{(j')}_{m'_j n'_j}(\varphi\theta\alpha) \frac{[(2\ell+1)(2\ell'+1)]^{\frac{1}{2}}}{4\pi} \sin\theta \, d\theta \, d\varphi \, d\alpha$$
$$= \sum_{LL'KN} \begin{pmatrix} \ell & j & L \\ -m & -m_j & -K \end{pmatrix} \begin{pmatrix} \ell & j & L \\ 0 & -n_j & -N \end{pmatrix} \begin{pmatrix} \ell' & j' & L' \\ m' & m' & K' \end{pmatrix} \begin{pmatrix} \ell' & j' & L' \\ 0 & n'_j & N' \end{pmatrix}$$
$$\delta_{LL'} \, \delta_{NN'} \, \delta_{MM'} \, \frac{[(2\ell+1)(2\ell'+1)]^{\frac{1}{2}}}{4\pi} \, \frac{(2L+1)(2L'+1)(-1)^{m+m_j-n_j-N+K}}{2L+1} \, 8\pi^2 \quad ,$$

or

$$= \sum_{KL} \begin{pmatrix} \ell & \ell' & L \\ -m & m' & -K \end{pmatrix} \begin{pmatrix} \ell & \ell' & L \\ 0 & 0 & 0 \end{pmatrix} \begin{pmatrix} j & j' & L' \\ -m_j & m'_j & K' \end{pmatrix} \begin{pmatrix} j & j' & L' \\ -n_j & n'_j & N' \end{pmatrix} \delta_{LL'} \delta_{N'0} \delta_{KK'}$$
$$[(2\ell+1)(2\ell'+1)]^{\frac{1}{2}} (2L+1) 2\pi (-1)^{m+m_j-n_j+k}$$

$$= \sum_{KL} \begin{pmatrix} \ell & j & L \\ -m & -m_j & -K \end{pmatrix} \begin{pmatrix} \ell & j & L \\ 0 & -n_j & n_j \end{pmatrix} \begin{pmatrix} \ell' & j' & L \\ m' & m_j & -K_j \end{pmatrix} \begin{pmatrix} \ell' & j' & L \\ 0 & n_j & -n_j \end{pmatrix}$$

$$[(2\ell+1)(2\ell'+1)]^{\frac{1}{2}}(2L+1)2\pi\delta_{n_j n_j'} \quad , \tag{2.8a}$$

or

$$\sum_{LK} \begin{pmatrix} \ell & \ell' & L \\ -m & m' & -K \end{pmatrix} \begin{pmatrix} \ell & \ell' & L \\ 0 & 0 & 0 \end{pmatrix} \begin{pmatrix} j & j' & L \\ -m_j & m_j' & K \end{pmatrix} \begin{pmatrix} j & j' & L \\ -n_j & n_j' & 0 \end{pmatrix} (-1)^{m+m_j-n_j+K}$$

$$[(2\ell+1)(2\ell'+1)]^{\frac{1}{2}}(2L+1)2\pi\delta_{n_j n_j'} \quad , \tag{2.8b}$$

giving

$$V^J_{ii}(R) = \sum_n [(2\ell+1)(2\ell'+1)]^{\frac{1}{2}} \begin{pmatrix} j & \ell & J \\ -n & 0 & n \end{pmatrix} \begin{pmatrix} j' & \ell' & J \\ -n & 0 & n \end{pmatrix} V^n_{\nu j \nu' j'}(R) \tag{2.9a}$$

or

$$= \sum_{nL} (-1)^{L+J+n}(2L+1) \begin{Bmatrix} j & \ell & J \\ \ell' & j' & L \end{Bmatrix} \begin{pmatrix} \ell & \ell' & L \\ 0 & 0 & 0 \end{pmatrix} \begin{pmatrix} j & j' & L \\ -n & n & 0 \end{pmatrix}$$

$$[(2\ell+1)(2\ell'+1)]^{\frac{1}{2}} \, V^n_{\nu j \nu' j'}(R) \quad . \tag{2.9b}$$

The form given by (2.9a) is convenient for the approximations which we will investigate later. The form given by (2.9b) looks somewhat more complex but simplifies drastically when the body-fixed potential is explicitly given. This is the form usually used in close coupling calculations. The body fixed potential integral may be written two different ways:

$$V^n_{\nu j \nu' j'}(R) = 2\pi \int_0^{\pi} \int_0^{2\pi} \varphi^{jn}_{\nu}(\vec{\eta}0\beta\gamma)V(0\beta\gamma 00R\vec{\eta})\varphi^{j'n}_{\nu'}(\vec{\eta}0\beta\gamma) \sin\beta \, d\beta \, d\gamma \, d\vec{\eta}$$

$$= \int_0^{2\pi} \int_0^{\pi} \int_0^{2\pi} \varphi^{jn}_{\nu}(\vec{\eta}\alpha\beta\gamma)V(\alpha\beta\gamma 00R\vec{\eta})\varphi^{j'n}_{\nu'}(\vec{\eta}\alpha\beta\gamma)\sin\beta \, d\alpha \, d\beta \, d\gamma \, d\vec{\eta} \quad , \tag{2.10}$$

the second being the most convenient as the integrals over the D matrices are well known.

2.1.2 Boundary Conditions

The boundary condition (2.2) is the physical boundary condition which defines the S-matrix. The S-matrix can be shown to be symmetric and unitary. In practice for most standard methods (in particular all of the methods of the NRCC tests), a real boundary condition given by

$$f^J_{i1}(R) \underset{R\sim\infty}{\sim} \frac{1}{|k_i|^{\frac{1}{2}}} \left[\overline{\sin}\left(k_i R - \frac{\ell\pi}{2}\right) \delta_{iI} + \bar{K}^J_{iI}\ \overline{\cos}\left(k_i R - \frac{\ell\pi}{2}\right)\right] \quad , \tag{2.11}$$

is used. The $\bar{K}$-matrix is a real symmetric matrix. This matrix is sometimes referred to as a reactance matrix, but that term is also often used for another matrix. The S-matrix is given in terms of this matrix by

$$\underline{S} = [\underline{1} + i\underline{\bar{K}}][\underline{1} - i\underline{\bar{K}}]^{-1} \quad . \tag{2.12}$$

The problem of closed channels arises in modern calculations. Though only the open channels of the S matrix are important for physical observables, the closed channel part of the $\bar{K}$ matrix in (2.12) may not be neglected. When i is a closed channel, k_i is pure imaginary. The $\overline{\sin}$ notation in (2.11) indicates that $\overline{\sin}\ (k_i R - \ell\pi/2)$ is to be replaced by $\sinh(|k_i|R - \ell\pi/2)$ when i is closed. A more convenient form to use in (2.11) for close channels is the S-matrix boundary conditions. This gives a modified K matrix solution to (2.11) which we will denote by a $\underline{K}$ without a bar. When this form is used, the S matrix may be computed for open channels using only the open channel K matrix. This is easily demonstrated by writing the f matrix in partitioned form, separating open and closed channels:

$$\begin{aligned}
\begin{pmatrix} \underline{f}_{oo} & \underline{f}_{oc} \\ \underline{f}_{co} & \underline{f}_{cc} \end{pmatrix} &= \begin{pmatrix} \underline{k}_i^{\frac{1}{2}} & \underline{0} \\ \underline{0} & |\underline{k}_i|^{\frac{1}{2}} \end{pmatrix} \left[\begin{pmatrix} \underline{\sin}\ \underline{kR} & \underline{0} \\ \underline{0} & \underline{e}^{kR} \end{pmatrix} \right. \\
&\quad \left. + \begin{pmatrix} \underline{\cos}\ \underline{kR} & \underline{0} \\ \underline{0} & \underline{e}^{-kR} \end{pmatrix} \begin{pmatrix} \underline{K}_{oo} & \underline{K}_{oc} \\ \underline{K}_{co} & \underline{K}_{cc} \end{pmatrix} \right] \\
&= \begin{pmatrix} \underline{k}_i^{\frac{1}{2}} & \underline{0} \\ \underline{0} & |\underline{k}_i|^{\frac{1}{2}} \end{pmatrix} \left[\begin{pmatrix} \underline{e}^{-ikR} & \underline{0} \\ \underline{0} & \underline{e}^{kR}\underline{P}^{-1} \end{pmatrix} \begin{pmatrix} \frac{i\underline{1}+\underline{K}_{oo}}{2} & \frac{\underline{k}_{oc}}{2} \\ \underline{0} & \underline{P} \end{pmatrix} \right. \\
&\quad \left. + \begin{pmatrix} \underline{e}^{ikR} & \underline{0} \\ \underline{0} & \underline{e}^{-kR}\underline{P} \end{pmatrix} \begin{pmatrix} -\frac{i}{2}\underline{1} + \frac{1}{2}\underline{K}_{oo} & \frac{1}{2}\underline{K}_{oc} \\ \underline{P}^{-1}\underline{K}_{co} & \underline{P}^{-1}\underline{K}_{cc} \end{pmatrix} \right] \quad .
\end{aligned} \tag{2.13}$$

Here, $\underline{\sin}\ \underline{kR}$ is the diagonal matrix with matrix elements $\sin(k_i R + \ell_i\pi/2)$ and similar elements for $\underline{\cos}\ \underline{kR}$. The matrix $\underline{\exp}(\pm kR)$ is diagonal with elements $\exp(\pm|k_i|R)$. For the $\underline{\sin}\ \underline{kR}$ and $\underline{\cos}\ \underline{kR}$ matrices, the i are open channels and for $\underline{\exp}(\pm kR)$ the i are closed channels. The matrix $\underline{P}$ is a matrix of phases. It is a diagonal matrix with elements $\exp(i\ell\pi/2)$. The matrices with oo subscript are those for which both initial and final states are open, oc have initial states closed and final open, etc.

It is easily shown that

$$\begin{pmatrix} \frac{i\underline{1}+\underline{K}_{oo}}{2} & \frac{\underline{K}_{oc}}{2} \\ \underline{0} & \underline{P} \end{pmatrix}^{-1} = \begin{pmatrix} \left(\frac{i\underline{1}+\underline{K}_{oo}}{2}\right)^{-1} & \underline{X} \\ \underline{0} & \underline{P}^{-1} \end{pmatrix} , \quad (2.14)$$

where $\underline{X}$ is a matrix product of simple form which need not concern us here. Multiplying (2.13) from the right by the matrix in (2.14), we easily see that

$$\begin{aligned} \underline{S}_{oo} &= -(-i\underline{1} + \underline{K}_{oo})(i\underline{1} + \underline{K}_{oo})^{-1} \\ &= (\underline{1} + i\underline{K}_{oo})(\underline{1} - i\underline{K}_{oo})^{-1} \quad . \end{aligned} \quad (2.15)$$

The boundary conditions of (2.13) are as easy to use as the usual K matrix conditions, if not easier. For long-ranged potentials, one may often drop closed channels before reaching the asymptotic region in any event, and the choice of boundary condition is academic. But in those cases for which one may stop the integration before the close channels have decayed away, the present scheme leads to a smaller complex matrix problem.

2.1.3 Asymptotic Approximations

The interaction potential can take on many forms. The long-range part of the potential, however, usually consists of a dispersion-multipole expansion in integral powers of 1/R . When the coupled equations are integrated into the region in which the potential is of this form, one may compute sets of solutions to the equations which obey the appropriate boundary conditions. These solutions are in the form of asymptotic expansions. The coefficients in the expansions are easy to compute by simple recursion relations [2.23-26] and [Ref.2.6, p.377]. The numerical wave function may be fit to the asymptotic series and the K matrix computed. This technique often allows one to terminate the integration of the coupled system soon after emerging from the attractive well. This approach allows one to terminate numerical integration as soon as the potential loses its rich structure. It is not only orders of magnitude more efficient than integrating into the asymptotic region, but it avoids the build up of rounding errors encountered in long integrations. It is particularly useful for problems such as charge-dipole collisions which have extremely long-range interactions. In a Li^+-N_2 scattering calculation performed [2.23] using an asymptotic expansion to terminate the integration, it was possible to use only a fraction of the integration range which would be required without this technique. Furthermore, a larger step size was possible for the integration. When the asymptotic equation was not used and the integration was carried numerically into the asymptotic region, there was a build up of rounding error due to truncation error in the integration formula because of the large number of steps needed. Thus we were required to use smaller steps when integrating numerically to the asymptotic region to achieve the same accuracy we obtained with the asymptotic formula.

The asymptotic series solutions are not convergent series and are only useful at R large enough that a finite number of terms gives sufficient accuracy. This may be explored experimentally by computing the K matrix at some large R and then continuing the numerical solution perhaps another DeBroglie wavelength or so and recomputing the K matrix. When two successive K matrices agree to the desired accuracy one may consider the integration complete. Such tests are easy to build into the computer code.

Long-range potentials often require very large total J quantum numbers to give fully converged elastic cross sections. For J large enough, both rotational and vibrational inelastic transitions become negligible and, when we are only interested in transitions out of j = 0, a single ordinary differential equation may be solved. This can be done using the WKB approximation, but even this is some effort. If we are interested in transitions from higher rotation states, we must solve a system coupled over ℓ transitions even when inelastic transitions are negligible. For very large J, the Born approximation works well if the K matrix form of the approximation is used. For large J, only the long-range $1/R^n$ part of the potential need be included in the calculation. The wave function may be written in the Lippman-Schwinger form

$$f_{iI}^{(J)}(R) = \delta_{iI}u_\ell(k_iR) + \int_0^\infty \frac{u_\ell(k_iR_<)v_\ell(k_iR_>)}{k_i} \sum_{i'} V_{ii'}^J(R)\, f_{i'I}^J(R)\, dR \quad , \tag{2.16}$$

where u_ℓ and v_ℓ are Riccati Bessel and Neuman functions, respectively:

$$u_\ell(x) = xj_\ell(x) \tag{2.17}$$

$$v_\ell(x) = xn_\ell(x) \quad . \tag{2.18}$$

In the Born approximation we replace $f_{i'I}^J$ on the right-hand side of (2.16) by $u_\ell(k_IR)$. Then the energetically elastic radial function is given by

$$f_{II'}^J(R) = u_\ell(k_IR) + \int \frac{u_\ell'(k_IR_<)v'_\ell(k_IR_>)}{k_I} V_{II'}^J(R)u_\ell'(k_I'R)\, dR \quad . \tag{2.19}$$

From the asymptotic form of $f_{II'}^J$ we obtain the K matrix

$$K_{II'}^J = \int_0^\infty \frac{u_\ell(k_IR)V_{II'}^J(R)u_{\ell'}(k_IR)}{k_I}\, dR \quad . \tag{2.20}$$

For J large, only the long-range part of the interaction potential contributes significantly to this integral. If the long-range part of the interaction potential is of the form

$$V^J_{I'I} = \sum_n \frac{a^J_{I'In}}{R^n} \tag{2.21}$$

as it often is, then the integral in (2.20) may be performed in closed form to give

$$K^J_{I'I} = \sum_n \frac{\pi}{2} a^J_{I'In} \frac{k^{n-1}(n-2)!\Gamma[(\ell+\ell'-n+3)/2]}{2^{n-1}\Gamma[(\ell-\ell'+n)/2]\Gamma[(\ell'-\ell+n)/2]\Gamma[(\ell+\ell'+n+1)/2]} \tag{2.22}$$

for $\ell + \ell' + 3 > n$. Two of the Γ functions in (2.22) reduce to factorials and the other two are Gamma functions of half integral argument which are easy to evaluate. Equation (2.22) is for energetically elastic K matrices; that is to say, for $k_I = k_{I'}$. The initial and final states differ only in ℓ. The inelastic integrals may also be performed using

$$\int_0^\infty \frac{u_\ell(kR)u_{\ell'}(k'R)}{R^n}\, dR$$

$$= \frac{\pi}{2} \frac{k^\ell\Gamma[(\ell+\ell'-n+3)/2]\,{}_2F_1[(\ell+\ell'-n+3)/2],[(\ell-\ell'-n+2)/2];\ell+(3/2);(k/k)^2}{2^{n+1}k'^{\ell-n+1}\Gamma[(\ell'-\ell+n)/2]\Gamma[\ell+(3/2)]} \tag{2.23}$$

for $k' > k$ and $\ell + \ell' + 3 > n$. If $k' < k$, then k and k' and ℓ and ℓ' must be interchanged in (2.23). ${}_2F_1$ is a hypergeometric function. This function does not converge for $k/k' = 1$ and (2.22) may be used in this special case. It is possible that the hypergeometric function may be difficult to evaluate when k/k' is close to one. I have never used (2.23) because in all cases for which I have used the Born approximation, the ℓ quantum numbers have been so large that only the elastic transitions are significant. I include (2.23) here only for completeness. It is certainly possible that for a charge-dipole or some other very long-range anisotropy, the Born approximation may be accurate for ℓ's for which inelastic transitions are still nonnegligible. For large ℓ it is only necessary to carry the elastic channels and the number of coupled equations are small. Nonetheless, it is much easier to evaluate (2.23) than to solve the coupled equations. For high energy long-range collisions, the Born approximation is accurate and so many J states are requited that even the WKB method becomes prohibitive.

In a calculation [2.23] of Li^+-N_2 collisions at 4.23 eV center of mass energy, it was necessary to consider angular momentum states up to J = 10,000 to obtain converged elastic cross sections. The inelastic transitions were negligible well before J = 1500 and we were able to use the Born approximation for J from 1500 to 10,000. This entire calculation required only a few seconds to compute all J states. Even using the WKB approximation the computation of the solution for this number of J states would have been impossible.

2.2 A Higher-Order Coupled States Approximation

For molecules which do not contain hydrogen, close coupling calculations become prohibitively expensive except at extremely low energies. This is especially true if one is interested in vibrational transitions. The number of rotation states which must be included in an accurate calculation becomes astronomical for heavy molecules at energies where vibrational transitions are important. The coupled states and infinite-order sudden approximations often bring these problems into the realm of feasibility. These approximations are excellent for some problem. For others the results are marginal, and for still other problems these approximations fail altogether. In every case, however, they appear to improve as the energy becomes larger. As we increase the energy we quickly reach the point at which it is no longer possible to do exact calculations. At this point we continue to use the approximations and hope that they go on improving as the energy gets higher.

We may, on the other hand, use these approximations as a zeroth-order approximation and improve them using perturbation techniques.

The angular momentum decoupling approximations have been discussed at great length in the literature [2.27] and we will give here only a sketch of the derivation to fix the notation and to present the working equations. We will then give a short discussion of the S-matrix phase and the transformations between the various representations. There has been much confusion over the phase problem in the literature and it is worth some space here to clear it up.

Let us discuss in detail the coupled states approximation. The points we make about coupled states will be applicable in an obvious way to other angular momentum decoupling approximations.

There are a number of different ways to derive the angular momentum decoupling approximations. An approach which applies to all angular momentum decoupling approximations is to diagonalize the potential matrix $V^J_{ii'}$ of (2.1) in the quantum number we wish to eliminate. For the coupled states approximation this quantum number is ℓ. We wish to find a transformation which will transform to a representation which does not depend on ℓ. This has been done a number of different ways in the literature [2.29]. From (2.9a), or (2.9b) we can see that it would not be hard to find a transformation to transform $V_{ii'}$ to $V^n_{\nu j\nu' j'}$. In fact, (2.9a,b) are inverses of such a transformation. The transformation can easily be found and was first given explicitly by HUNTER [2.28] but it is not necessary to know this transformation. We need only note that if the transformation, whatever it is, is applied to (2.1), it will do unpleasant things to the $\ell(\ell + 1)/R^2$ term. To avoid this problem we replace $\ell(\ell + 1)$ on the left of (2.1) by $\mathscr{L}^2$, some number independent of ℓ. This is the coupled states approximation. We then have the equation

$$\left(-\frac{\partial^2}{\partial R^2}+\frac{\mathscr{L}^2}{R^2}-k_i^2\right)g^n_{iI}(R) = -\sum_{i'} V^n_{\nu j\nu' j'}\, g^n_{i'I} \tag{2.24}$$

where now the i subscript contains only ν and j quantum numbers. Using the reverse transformation given in (2.9a), we obtain

$$\left(-\frac{\partial^2}{\partial R^2}+\frac{\mathscr{L}^2}{R^2}-k_i^2\right)f^J_{iI}(R) = -\sum_{i'} V^J_{\nu j\ell\nu' j'\ell'}\, f^J_{i'I} \quad , \tag{2.25}$$

where

$$f^J_{j\nu\ell j_I\nu_I\ell_I}(R) = \sum_{nn_J} [(2\ell+1)(2\ell_I+1)]^{\frac{1}{2}} \begin{pmatrix} j & \ell & J \\ -n & 0 & n \end{pmatrix} \begin{pmatrix} j_I & \ell_I & J \\ -n_I & 0 & n_I \end{pmatrix} \delta_{nn_I} g^n_{j\nu j_I\nu_I}(R) \quad . \tag{2.26}$$

Thus we have an exact solution to (2.25) in terms of the solution of the simpler (2.24). Equation (2.25) is the coupled states equation, differing from the close coupling equation only by replacing $\ell(\ell + 1)$ on the left by $\mathscr{L}^2$. The difference between the various coupled states approximations consists of making different choices of $\mathscr{L}^2$. This is all well known and is in the literature. It will be noticed that we made no use of the back transformation, i.e., the transformation from f to g. This transformation is very easy to find [2.28], however, and when one applies it to the close coupling equation (2.1), one obtains

$$\left(-\frac{\partial^2}{\partial R^2}+\frac{J(J+1)+j(j+1)-2n^2}{R^2}\right) g^{nn_I}_{j\nu' j_I\nu_I}$$

$$- k^2_{j\nu} + [(J-n+1)(J+n)(j-n+1)(j+n)]^{\frac{1}{2}} g^{n-1n_I}_{j\nu j_I\nu_I}$$

$$+ [(J+n+1)(J-n)(j+n+1)(j-n)]^{\frac{1}{2}} g^{n+1n_I}_{j\nu j_I\nu_I} = -\sum_{i'} V^n_{j\nu j'\nu'}(R)\, g^{nn_I}_{j\nu j_I\nu_I}(R) \quad . \tag{2.27}$$

This is just the close coupling equations in the body fixed coordinate system. We can see from this that the coupled states approximation consists of neglecting the two terms off-diagonal in n and replacing the $J(J+1)+j(j+1)-2n^2$ term by $\mathscr{L}^2$. The approximation one obtains by simply neglecting the off-diagonal terms and leaving the diagonal term untouched was studied by TAMIR and SHAPIRO [2.29] and found to be not quite as good as some of the other coupled states approximations [2.30]. This approximation has not been investigated in detail, however.

2.2.1 The Phase in the Body Fixed Coordinate System

Now we are ready to discuss the phase problem. In the space fixed coordinate system, the traditional boundary condition of (2.1) is given by (2.2). The question which has been much discussed in the literature is what is the "proper" boundary condition in the body fixed coordinate system. What is the boundary condition one should use for (2.27)? A number of workers have come up with different answers to this question. It is not a question which can be answered in isolation and all of these workers could be right if they treat their result properly. It would serve no purpose to review all of the answers to this question, but let me discuss one solution to the problem which has been proposed.

One approach [2.31] to the solution is to take the boundary condition in the space fixed coordinate system and transform it to the body fixed coordinate system, using the same transformation used for transforming the space fixed equation to the body fixed system. This is a reasonable and almost obvious approach. It leads, however, to rather messy boundary conditions in the body fixed system.

In practice, when one solves the coupled system of equations one finds a complete set of solutions. Thus it does not really matter what boundary conditions are used in the body fixed coordinate system, if one then transforms back to the space fixed system properly. When one does that he will not in general obtain the boundary conditions of (2.2), however. If a complete set is transformed to the space fixed system, proper linear combinations may be found which give the desired boundary condition in the space fixed system. A simple and straightforward solution to this problem can be found. We choose a boundary condition in the body fixed system which transforms to a simple form in the space fixed system. The boundary condition we use with (2.27) is

$$g^{nn_I}_{j\nu j_I\nu_I} \sim \frac{1}{k^{1/2}_{j\nu}}\left[\exp(-ik_{j_I\nu_I}R)\delta_{jj_I}\delta_{\nu\nu_I}\delta_{nn_I} - S^{nn_I}_{j\nu j_I\nu_I}\exp(ik_{j\nu}R)\right] \quad . \tag{2.28}$$

Transforming this solution to the space fixed system by the transformation of (2.26) gives

$$\bar{f}_{j\nu\ell j_I\nu_I\ell_I} \sim \frac{1}{k^{1/2}_{j\nu}}\left[\exp(-ik_{j\nu}R)\delta_{jj_I}\delta_{\nu\nu_I}\delta_{\ell\ell_I}\right.$$
$$\left. - \exp(ik_{j\nu}R)\sum_{nn_I}[(2\ell+1)(2\ell_I+1)]^{1/2}\begin{pmatrix} j & \ell & J \\ -n & 0 & n\end{pmatrix}\begin{pmatrix} j_I & \ell_I & J \\ -n_I & 0 & n_I\end{pmatrix} S^{nn_I}_{\nu j\nu_I j_I}\right] \quad . \tag{2.29}$$

Then from (2.2) we see that

$$f^J_{j\nu\ell j_I\nu_I\ell_I} = \exp(\ell_I\pi/2)\,\bar{f}^J_{j\nu\ell j_I\nu_I\ell_I} \tag{2.30}$$

and

$$S^J_{\nu j\ell\nu_I j_I\ell_I} = \exp\left[\frac{(\ell_I+\ell)\pi}{2}\right]\sum_{nn_I}[(2\ell+1)(2\ell_I+1)]^{1/2}$$

$$\begin{pmatrix} j & \ell & J \\ -n & 0 & n \end{pmatrix} \begin{pmatrix} j_I & \ell_I & J \\ -n_I & 0 & n_I \end{pmatrix} S^{nn_I}_{\nu j \nu_I j_I} \quad . \tag{2.31}$$

The same phase convention may be used with the coupled states equations, (2.24). Any phase in (2.28) would serve as well if one is careful to transform to the space fixed system properly and then correct the phase of the solution obtained. The phase used in (2.28) is particularly simple and leads to the extremely simple expression for the space fixed S matrix (2.31).

In the coupled states approximation, of course, since we have neglected the off-diagonal elements in n, we have

$$S^{nn_I CS}_{\nu j \nu_I j_I} = S^{n}_{\nu j \nu_I j_I} \delta_{nn_I} \quad . \tag{2.32}$$

2.2.2 Correcting the Coupled States Approximation

The coupled states approximation gives a wave function close to the correct one. One is thus led to consider improvement of the solution by perturbation theory. In the usual distorted wave theory one uses for the unperturbed solution the equation obtained by neglecting all off-diagonal terms in the potential matrix. With such a reference Hamiltonian the unperturbed solution gives only elastic scattering. All of the inelasticity is treated by perturbation. Even so the results are often reasonable. The use of the coupled states solution as the unperturbed solution offers great improvements over the distorted wave method. All of the interaction potential is included in the unperturbed potential. The perturbation in the body fixed coordinate system consists only of the off-diagonal terms on the left of (2.27) and the diagonal perturbation

$$V' = \frac{J(J+1) + j(j+1) - 2n^2 - \mathscr{L}^2}{R^2} \quad . \tag{2.33}$$

Of course $\mathscr{L}^2$ may be chosen so that V' is zero. This corresponds to the coupled states solution of TAMIR and SHAPIRO [2.29]. Though this may not be the best choice of reference Hamiltonian [2.29], preliminary calculations [2.20] have shown that even with this approximation the perturbation corrections are excellent. With better coupled states approximation, the results are even more impressive.

There are two problems encountered in using a perturbation approach to correct the coupled states equations. First, the solution is not a simple function but a matrix solution. Thus, to develop a perturbation correction one needs a matrix Green's function for the problem. Second, one must compute integrals of the perturbation function with the solutions to the coupled states equations. The methods currently in use give only the asymptotic form of the solution [Ref.2.6, p.265].

The invariant embedding techniques never compute the solution to the entire problem but compute solutions to intermediate problems which become the solution to the correct problem only in the asymptotic region. The solution-following techniques do compute the solution everywhere, but these techniques are unstable. Various stabilizing methods must be employed so that the correct asymptotic form of the solution will be preserved in the asymptotic region. One might feel that these stabilizing transformations could be saved and then, when the proper asymptotic form of the solution is found, the transformations could be reversed to give the solution everywhere. This approach does not work, however. The stabilizing transformations are built to stabilize as one proceeds from the nonclassical region into the asymptotic region. When they are reversed they become destabilizing transformations and rounding error builds drastically.

We have managed to overcome both of these problems [2.20]. The first was simple. The matrix Green's function was rather easy to derive and since our derivation it has been derived by three other workers independently [2.32-34]. As for a Green's function for an ordinary differential equation, the matrix Green's function is not unique but depends on the boundary conditions desired for the solution one wishes to obtain. Since the derivation of matrix Green's functions is in the literature [2.32,33,35], we will give only the final result here.

The equation we wish to solve has the form

$$\left[\frac{\partial^2}{\partial x^2} + \underline{V}(x)\right]\underline{f}(x) = \underline{W}(x) \quad , \tag{2.34}$$

where $\underline{V}$, $\underline{W}$ and $\underline{f}$ are matrix functions of x. Assume that we know solutions to the homogeneous equation

$$\left[\frac{\partial^2}{\partial x^2} + \underline{V}(x)\right]\begin{pmatrix}\underline{u}(x)\\ \underline{v}(x)\end{pmatrix} = 0 \tag{2.35}$$

such that $\underline{u}(x)$ satisfies the desired boundary condition at one boundary and $\underline{v}(x)$ satisfies the boundary condition at the other boundary. Then we may write the solution $\underline{f}(x)$ of (2.34) as an integral

$$\underline{f}(x) = \underline{u}(x) + \int_{x_0}^{x} \underline{G}(x\ x')\underline{W}(x')d\,x' \quad , \tag{2.36}$$

where $\underline{u}(x)$ is the solution of (2.35) which satisfies the boundary condition at the boundary x_0. The matrix Green's function $\underline{G}$ is given by

$$\begin{aligned}\underline{G}(x,x') &- \underline{w}^{-1}\underline{u}(x)\underline{v}^T(x') \qquad x < x' \\ &\underline{w}^{-1}\underline{v}(x)\underline{u}^T(x') \quad , \quad x > x'\end{aligned} \tag{2.37}$$

where $\underline{w}$ is a diagonal Wronskian matrix [2.35]. This Green's function is in many ways similar to a Green's function for uncoupled systems, though the proof that it solves the equations is somewhat involved [2.35].

For the coupled states apprxomation we write the coupled radial equations in matrix form:

$$\left(-\frac{\partial^2}{\partial R^2} + \frac{\mathscr{L}^2}{R^2} - \underline{k}^2\right)\underline{g} + \underline{V}\underline{g} = -\frac{L^2 - \mathscr{L}^2}{R^2}\underline{g} - \frac{\Omega}{R^2}\underline{g} \tag{2.38}$$

where $\mathscr{L}^2$ is the diagonal approximate relative angular momentum, $\underline{k}^2$ is the diagonal kinetic energy matrix, L^2 is the exact diagonal angular momentum in the body fixed coordinate system, $J(J+1) + j(j+1) - 2n^2$, and Ω is the off-diagonal matrix given explicitly in (2.27). This equation (2.38) is exact and is just a rearrangement of (2.27) written in matrix form. The solution using the matrix Green's function is then

$$\underline{g} = \underline{u} - \int_0^\infty \underline{G}(R,R')\left(\frac{L^2 - \mathscr{L}^2}{R'^2} + \frac{\Omega}{R'^2}\right)\underline{g}(R')dR' \quad , \tag{2.39}$$

where $\underline{u}$ is the matrix of coupled states solutions, i.e., $\underline{u}$ is the solution of (2.38) with the right-hand side set equal to zero. It is regular at the origin and we take its asymptotic form to be

$$(\underline{u})_{\nu j \nu_I j_I} \underset{R\to\infty}{\sim} \delta_{nn_I} k_{\nu j}^{-\frac{1}{2}}(\delta_{jj_I}\delta_{\nu\nu_I} \sin k_{\nu j}R + K^n_{\nu j \nu_I j_I} \cos k_{\nu j}R) \quad . \tag{2.40}$$

The $\underline{v}$ (2.37) in the Green's function is the solution irregular at the origin with asymptotic form

$$(\underline{v})_{\nu j \nu_I j_I} \sim k_{\nu j}^{-\frac{1}{2}}\delta_{\nu\nu I}\delta_{jj_I}\delta_{nn_I} \cos k_{\nu j}R \quad . \tag{2.41}$$

For these boundary conditions the Wronskian $\underline{w}$ is unity. We note that (2.39) is exact.

We now make an approximation similar in spirit to the distorted wave approximation. We replace $\underline{g}$ in the integral of (2.39) by $\underline{u}$, the unperturbed coupled states wave function matrix. If we now look at the asymptotic form of our approximate $\underline{g}$, using (2.37) for the Green's function we obtain

$$\underline{g}^c \underset{R\to\infty}{\sim} \underline{u} - \underline{v}\int_0^\infty \underline{u}^T \frac{L^2 - \mathscr{L}^2 + \Omega}{R'^2}\underline{u}\, dR'$$

$$= k^{-\frac{1}{2}}(\sin \underline{k}R + \cos kR\underline{K}^{cs})$$

$$- \cos kR \int_0^\infty \underline{u}^T \frac{L^2 - \mathscr{L}^2 + \Omega}{R'^2} \underline{u}\, dR') \quad . \tag{2.42}$$

From this analysis we get the corrected K-matrix

$$\underline{K}^c = \underline{K}^{cs} - \int_0^\infty \underline{u}^T \frac{L^2 - \mathscr{L}^2 + \Omega}{R'^2} \underline{u}\, dR' \quad . \tag{2.43}$$

We notice that we need only the regular solution $\underline{u}$ to the coupled states equations. The asymptotic form of the irregular solution $\underline{v}$ served to evaluate the Wronskian which occured in the Green's function and we need know no more about it.

The phase of the solution in the body fixed coordinate system will, of course, have some effect on the correction to $\underline{K}^{cs}$. As we have shown, if the phase is corrected after transformation back to the space fixed coordinate system, then it makes no difference which phase we use in the body fixed coordinate system for simple CS calculations, but when corrections are made in the manner described in (2.43), there will be a difference which depends on the phase chosen. We have made preliminary calculations using different phase choices [2.35]. On the basis of the one problem we have solved so far, there is no clear choice of phase. All choices gave a significant improvement over the uncorrected CS calculation and for every phase some matrix elements were better than for any of the other choices, but the variation in the results were minor.

The infinite-order sudden (IOS) approximation [2.8,10] is a further approximation in which one makes both the centrifugal sudden approximation and the energy sudden approximation. There are many situations for which the energy sudden approximation would be an excellent approximation and the centrifugal sudden not so good. The present approach to correcting the centrifugal sudden part of the IOS approximation would be particularly easy to apply. This would lead to a particuarly simple approximation for high energy collisions. It would have the added advantage of allowing us to estimate the magnitude of the error committed in using the centrifugal sudden approximations at higher energy.

References

2.1 T.F. George, J. Ross: Ann. Rev. Phys. Chem. *24*, 263 (1973)
2.2 D. Secrest: Ann. Rev. Phys. Chem. *24*, 379 (1973)
2.3 H. Rabitz: In *Modern Theoretical Chemistry*, ed. by W.H. Miller (Plenum Press, New York 1976)
2.4 W.H. Miller (ed): *Dynamics of Molecular Collisions* (Plenum Press, New York 1976)
2.5 J.P. Toennies: Ann. Rev. Phys. Chem. *27*, 225 (1976)
2.6 R.B. Bernstein (ed.): *Atom-Molecule Collision Theory. A Guide for the Experimentalist* (Plenum Press, New York 1979)
2.7 H. Rabitz: J. Chem. Phys. *57*, 1718 (1972); see also [2.3]
2.8 D.J. Kouri: [Ref.2.6, p.301]
P. McGuire, D.J. Kouri: J. Chem. Phys. *60*, 2488 (1974)

2.9 R.B. Walker, J.C. Light: Chem. Phys. *7*, 84 (1975)
2.10 T.P. Tsien, G.A. Parker, R.T. Pack: J. Chem. Phys. *59*, 5373 (1973)
2.11 D. Secrest: J. Chem. Phys. *62*, 710 (1975)
2.12 V. Khare: J. Chem. Phys. *68*, 4631 (1978)
2.13 L.D. Thomas (ed.): *Proceedings of the NRCC Workshop on Algorithms and Computer Codes in Atomic and Molecular Scattering*, Vol. I LBL-9501 (1979)
2.14 L.D. Thomas (ed.): *Proceedings of the NRCC Workshop on Algorithms and Computer Codes in Atomic and Molecular Quantum Scattering*, Vol. II LBL-9501 (1980)
2.15 L.D. Thomas, M.H. Alexander, B.R. Johnson, W.A. Lester Jr., J.C. Light, K.D. McLenithan, G.A. Parker, M.J. Redmon, T.G. Schmalz, D. Secrest, R.B. Walker: J. Comp. Phys. *41*, 407 (1981)
2.16 D. Secrest: *Methods of Computational Physics*, Vol. 10, ed. by B. Alder, S. Fernbach, M. Rotenberg (Academic Press, New York 1971) p.243
2.17 G.A. Parker, J.C. Light, B.R. Johnson: Chem. Phys. Lett. *73*, 572 (1980)
2.18 B.R. Johnson: J. Comp. Phys. *13*, 445 (1973)
2.19 G.A. Parker, T.G. Schmalz, J.C. Light: J. Chem. Phys. *73*, 1757 (1980)
2.20 K.D. McLenithan, D. Secrest: J. Chem. Phys. *73*, 2513 (1980)
2.21 D. Estes, D. Secrest: To appear
2.22 A.R. Edmonds: *Angular Momentum in Quantum Mechanics*, 2nd ed. (Univ. Press, Princeton, N.J. 1974)
2.23 G.A. Pfeffer, D. Secrest: To appear
2.24 M.A. Brandt: Masters Thesis, University of Minnesota, Minneapolis, Minnesota (1975)
2.25 P.G. Burke, H.M. Schey: Phys. Rev. *126*, 147 (1962)
2.26 M.A. Brandt, D.G. Truhlar: Chem. Phys. Lett. *23*, 48 (1973)
2.27 D.J. Kouri: [2.6] p.301 and the references cited therein
2.28 L.W. Hunter: J. Chem. Phys. *62*, 2855 (1975)
2.29 M. Tamir, M. Shapiro: Chem. Phys. *13*, 215 (1976); Chem. Phys. Lett. *39*, 79 (1976)
2.30 L. Monchick, S. Green: J. Chem. Phys. *66*, 3085 (1976)
2.31 D.J. Kouri, Y. Shimoni: J. Chem. Phys. *65*, 5020 (1976)
2.32 J.-T. Hwang, E.P. Dougherty, S. Rabitz, H. Rabitz: J. Chem. Phys. *69*, 5180 (1978);
J.-T. Hwang, H. Rabitz: J. Chem. Phys. *70*, 4609 (1979)
2.33 R.K. Nesbet: *Variational Methods in Electron-Atom Scattering* (Plenum Press, New York 1980)
2.34 L.D. Thomas: Private communication
2.35 K.D. McLenithan, D. Secrest: To appear

3. Quasiclassical Trajectory Studies of State to State Collisional Energy Transfer in Polyatomic Molecules

G.C. Schatz

With 2 Figures

This chapter presents a detailed analysis of the quasiclassical trajectory method as applied to the calculation of state resolved collisional energy transfer cross sections and rate constants in polyatomic molecule collision systems. It begins (Sect.3.1) with a brief review of previous applications of this type. These applications show that a major difficulty with describing polyatomic molecule collisions using classical methods arises in the specification of initial and final molecular semiclassical eigenstates. Both anharmonic and Coriolis coupling effects cause the molecular Hamiltonian to be nonseparable, and often the calculated energy transfer rates and pathways can be significantly in error if these nonseparable effects are not included in the trajectory initial and final conditions. The proper way to define semiclassical eigenstates, via good action-angle variables, requires a difficult solution to the molecular Hamilton-Jacobi equation. Section 3.2 describes several ways to determine semiclassical eigenstates based on approximate partitionings of the vibration-rotation Hamiltonian (to simplify Coriolis effects) and a perturbation theory solution to the Hamilton-Jacobi equation for the good action-angle variables governing vibrational motions. Section 3.3 discusses the application of these methods for describing molecular internal states to trajectory studies of collisional energy transfer, focusing specifically on the transformations between cartesian and good variables and vica versa. Section 3.4 presents the results of an application of the methods of Sects.3.2,3 to collisional excitation in the He + SO_2 system. The final state distributions are characterized and comparisons of a rotational sudden treatment of the dynamics with a fully coupled treatment are made. Examining the final state distributions, we find that they are neither uncorrelated between different modes nor are they so strongly correlated that only the energy gap between the initial and final state is important. The rotational sudden approximation is found to describe state to state energy transfer in He + SO_2 at 1-2 eV translational energy adequately, with a factor of 2 reduction in computational effort.

These trajectory applications illustrate the current state of the art in the treatment of collisional energy transfer in polyatomic systems using classical methods. A number of problems with such applications are discussed, including the difficulties associated with final state assignment when transitions are classically forbidden, and the inability of perturbation theory to treat Fermi resonance effects simply. Generalizations to the treatment of highly rotationally excited molecules, to molecules whose vibrational motion is chaotic rather than quasiperiodic, and to molecules with more than 3 vibrational modes are indicated as important topics of future research.

3.1 Background

Although the quasiclassical trajectory method has become the true workhorse of theoretical chemical dynamics, it has been used in surprisingly few studies of collisional energy transfer involving molecules other than diatomics. Unlike many problems in chemical dynamics, the reasons for this are not simply computational in nature; indeed, some of our basic understanding of the relationship between classical and quantum mechanics is at the heart of the issue. It is the purpose of this paper to explore some of the problems associated with such applications and to describe some of the methods now being developed to solve them.

The motivations for using trajectory methods to study vibrational and rotational energy transfer in polyatomic systems are two-fold. First, there is a need for theoretical interpretation of the large number of high quality measurements of cross sections and rate constants for state resolved energy transfer in such systems. Second, the quasiclassical trajectory method provides a nonperturbative, versatile and potentially accurate approach for studying large molecule collision processes, and thus should give a description of such processes which is unencumbered by dynamical approximations. The energy transfer experiments range from crossed beam studies of collisional excitation in CO_2, N_2O, CF_4, SF_6 and other molecules by fast Li^+ and H^+ ions using an angular and state resolved time of flight analysis [3.1] to laser-induced fluorescence studies [3.2] of vibrational relaxation in molecules such as cyclopropane [3.2b] in bulb conditions. At present, the most commonly used theoretical methods available for studying these collisional energy transfer processes are SSH and related approximate methods [3.3] (we omit from this discussion the treatment of near resonant processes). These methods are easy to apply and have been parameterized to predict energy transfer rates which correlate well with experimental measurements. Their dynamical basis is weak, however, so that the physical significance of parameters derived from them is not always clear. The theoretical foundation of the quasiclassical trajectory method is by contrast well established via semiclassical S-matrix theory and numerous studies of energy transfer in diatomic systems [3.4] have proven the accuracy of trajectory methods in practical applications. It would, of course, be preferable to use accurate quantum methods for studying polyatomic collision processes, but such methods are at present still in their infancy [3.5] and are not yet routinely available for studying larger systems. The large number of states accessible in most polyatomic systems of interest will certainly limit the types of problems which can be treated by quantum methods. At the same time, classical methods are most accurate in treating systems with large numbers of accessible quantum states.

Despite the apparent simplicity of using trajectory methods to study polyatomic collisions, the few studies of this type which have been done have been fraught with difficulties. One of the earlier trajectory applications was by THOMMARSON et al. [3.6] who considered the excitation of $CO_2(000)$ to (100,010,001) in collisions with He (using trajectories integrated in two dimensions). Excitation probabilities were calculated by equating the classical and quantal energy transfer moments for each mode. Although this study was able to produce probabilities which correlated well with experimental results (or SSH theory), even using a very crude intermolecular potential, an important approximation was used in the study which makes the significance of the results uncertain. Specifically, the CO_2 force field was assumed to be a purely quadratic function of the internal coordinates, thereby neglecting anharmonic terms in the intramolecular potential. This makes the determination of the semiclassical eigenstates describing CO_2 vibrational motion quite easy (each normal mode is governed by an uncoupled harmonic oscillator potential), but this approximation omits Fermi resonant effects that cause the (020) and (100) states in CO_2 to be extensively mixed. Had the anharmonic terms in the potential been included, as was done by SUZUKAWA et al. [3.7] some years later in a study of Kr + CO_2, they would have noticed that the energies in the bend and symmetric stretch modes were not constants of the motion in the absence of collisional interaction. Rather, because of anharmonic coupling, the energy in the symmetric stretch constantly oscillates, flowing into the bend and back, as a classical manifestation of the Fermi resonance. Thus, the energy moments for each mode depend on when (after the collision) they are calculated, obviously not a desirable situation. The approach of SUZUKAWA et al. [3.7] to circumvent this was to time average the energies associated with each mode after each collision. This, however, was later shown to produce results of questionable reliability [3.8]. Another approach to circumvent this problem was used by SATHYMURTHY and RAFF [3.9] in a study of CO_2 bending excitation in collisions with H_2. They chose to give the CO_2 zero energy rather than zero point energy initially in each stretch mode. With this treatment, even with an anharmonic force field, Fermi-resonance effects are negligible and stationary normal mode energies are specified prior to each collision. SATHYMURTHY and RAFF found that the amount of collisionally induced stretch excitation was small so that stationary vibrational actions could also be calculated after each collision (with some time averaging). Although the validity of these approximations has never been tested, it is clear that this approach is inherently limited to certain types of initial and final molecular eigenstates (those having zero energy in all modes except one). Moreover, the existence of Fermi coupled states tells us that the very idea of trying to calculate uncoupled normal mode energies is physically unrealistic in many cases.

Because of these difficulties with the classical description of the internal states of polyatomic molecules, many classical studies of nonreactive collisions involving polyatomics have not attempted to extract mode-resolved information about the collisions. This was true in the studies of collisional energy transfer in O_3 and H_2O (using He, Ar and Xe as collision partners) by STACE and MURRELL [3.10], who chose to evaluate only the total internal energy change (vibration plus rotation) in the O_3 and H_2O. This study as well as those of [3.7,9] also addressed a related problem in the treatment of energy transfer in polyatomics, namely, how to separate rotational from vibrational energy in the polyatomic internal state description. The effect of the Coriolis coupling between vibration and rotation is to cause vibrational and rotational energies to oscillate in time. Sometimes such couplings are relatively small so the oscillations can either be neglected or time averaged with little difficulty. This was done in the CO_2 studies in [3.7,9]. When the rotational energy is comparable to or larger than the vibrational energy, the Coriolis effect can be so large that the separation of vibration and rotation becomes meaningless. This was the case for some of the systems studied by STACE and MURRELL since they chose the H_2O and O_3 initial conditions from a Boltzmann distribution at temperatures ranging from 500 to 10000 K (vibrational zero point energy was omitted). They circumvented this problem by analyzing only the total change in molecular internal energy, but in doing so they omitted much of the interesting information which could have been extracted from their trajectory results had some method for separation been available. They did attempt to elucidate some information about the partitioning of energy transfer between vibration and rotation by integrating some trajectory ensembles chosen to have zero initial rotational or vibrational energies. From these results, they concluded that much of the energy transfer into the H_2O and O_3 was rotational. A later study by MULLONEY and SCHATZ [3.11] found that rotational sudden approximations could be used to effect a reasonable partitioning, and the conclusions of [3.10] concerning the relative amounts of vibrational and rotational energy transfer were verified and quantified.

Other trajectory studies of nonreactive collisions involving polyatomics have concentrated on collision lifetime information rather than on mode resolved energy transfer. These include a model study of the Cl + CD_3 system by MACDONALD and MARCUS [3.12] and several studies of the H + C_2H_4 system by HASE and coworkers [3.13]. The lifetimes calculated in these studies were used to interpret theories of unimolecular rate processes which fall outside the scope of this paper. It is, however, relevant to note that the problems concerning initial condition specification described above (anharmonic and Coriolis effects) were also present in these studies and inhibited the evaluation of mode resolved energy transfer information in them. Indeed, in some of these studies, the molecules considered were in such high energy states that anharmonic terms in the intramolecular potential accounted for a large fraction of the available molecular energy. Thus, the commonly used procedure [3.14] of defining polyatomic molecule initial conditions using the harmonic Hamiltonian, then scaling the resulting momenta (so as to correct for anharmonic and Coriolis contributions to the total energy), represents a significant extrapolation from the uncoupled result and produces an initial phase space distribution which is probably not stationary.

Still another approach to the use of trajectory methods to determine collisional energy transfer information in polyatomic systems has been the semiclassical work of BILLING [3.15]. This is actually a mixed quantum/classical approach in which reduced dimensionality trajectories are integrated to determine the time evolution of the translational, rotational and perhaps certain vibrational degrees of freedom in the colliding molecules, then the time dependent Schrödinger equation is solved to determine transition probabilities for the remaining vibrational degrees of freedom. BILLING's method allows for inclusion of anharmonic and Coriolis effects in describing those degrees of freedom assumed to be governed by classical mechanics and (separately) for those governed by quantum mechanics. A partitioning between the degrees of freedom which are treated by classical mechanics and those by quantum mechanics is necessary, however, and no anharmonic or other coupling between the classical and quantum degrees of freedom may be included. Although BILLING has used this approach in an impressive number of applications to polyatomic systems [3.15], often with excellent results, the accuracy of the approximations he makes has not been clearly established. The advent of accurate quantum methods for studying collisional energy transfer in triatomics [3.5] should help clarify this.

The main focus of this paper concerns a series of trajectory studies of collisional energy transfer in polyatomics which has appeared recently by SCHATZ and coworkers [3.8,16,17] in which anharmonic and Coriolis effects are treated accurately in the determination of initial and final trajectory conditions, thereby overcoming the problems experienced in many of the earlier studies. This accurate treatment involves the determination of the classical constants of the motion, the so-called "good" action variables, which describe the vibration-rotation motions of each polyatomic. These good actions are obtained by solving the molecular Hamilton-Jacobi equation, and in the presence of Coriolis or anharmonic coupling, leads to a proper description of the mixing between different modes. In addition, the good actions are directly related using semiclassical theory to polyatomic molecule quantum numbers (the former divided by $\hbar$ equals the latter plus a zero point term), thus enabling the specification of molecular semiclassical eigenstates. Thus, by determining these good action variables (and their conjugate angle variables), both the initial and final conditions needed in quasiclassical trajectory calculations can be specified, thereby enabling the study of state to state processes involving polyatomic molecules.

The first use of good action-angle variables in studies of collision processes was in an application to collisional energy transfer in the collinear Kr + CO_2 system [3.8,16]. In this work, the influence of anharmonic terms in the intramolecular potential on energy transfer processes was studied and found to be quite significant. At low collision energies, the primary effect of the anharmonic terms was to facilitate V → V processes, while at high collision energies, the V → T rates were also strong influenced by them. This first study used a method developed by CHAPMAN et al. [3.18] to determine action-angle variables. In a later paper [3.17] a much simpler but more approximate method based on classical perturbation theory was utilized to study collinear Kr + CO_2 and was found to be adequate for most transitions of interest. This perturbation method has since then been applied to the study of collisional excitation in Li^+ + CO_2 [3.19] (using a three-dimensional rotational sudden method), and in the O + H_2O [3.20] and O + CO_2 [3.21] systems (using both the sudden and full 3D dynamics). Applications to state to state reactive scattering in the systems O + CS_2 → SO + CS [3.22] and OH + H_2 → H_2O + H [3.23] have also been considered.

The theoretical methods for determining good action-angle variables and defining semiclassical eigenstates used in these applications date back to studies originally done in the days of the Old Quantum Theory [3.24]. There has been a resurgence of interest lately in theories of good action variables since an understanding of the polyatomic molecular vibrational motions which they describe is crucial to the development of quantitative theories of unimolecular reactions, radiationless transitions, multiphoton dissociation and many other processes. Since there are now many excellent reviews of both the older and more recent literature on this topic [3.25], we refer the reader to them for a detailed discussion of the basic theories involved. For the purposes of this discussion, we need only consider two relevant questions: (a) When do good action-angle variables exist? (b) If they exist, how are they accurately and efficiently determined?

The question of existence of good action-angle variables is difficult to answer, and in fact even though it was pondered by astronomers long ago [3.26], a clear cut solution still has not appeared. It seems apparent, however, that their existence is at least related to the regularity of molecular vibrational motion. If this motion is regular or quasiperiodic, then 3N-5 good action variables can be defined (including one for rotation), while if the motion is irregular or chaotic, constants of the motion other than total energy, linear and angular momenta do not exist. For most molecules, trajectory motion tends to be chaotic at molecular energies close to the dissociation energy and quasiperiodic at lower energies. The transition from regular to chaotic behavior as the energy increases is sometimes fairly abrupt, and appears to correlate with a change from regular to chaotic behavior in the quantum mechanical spectrum of the molecule.

Although good actions do not rigorously exist in the chaotic regime, JAFFE and REINHARDT [3.27] have shown how to define them approximately at energies all the way to the dissociative limit. Whether such methods will be needed in studies of collisions involving highly excited molecules is a bit uncertain at present since the importance of mode specificity in the chaotic regime has been the subject of only one model study [3.28]. This was a model of a collinear atom-triatom collision system, wherein the triatomic intramolecular potential was taken as the well studied Henon-Heiles potential [3.25]. The object of this study was to examine the variation of $V \rightarrow T$ energy transfer moments with initial triatomic conditions for a given translational energy. The initial conditions were defined by selecting points at random from a nearly periodic trajectory describing molecular vibrational motions at a given energy. Several different ensembles were generated using different nearly periodic trajectories, and at low molecular energies where molecular trajectory motion was quasiperiodic, the energy transfer varied quite substantially with initial conditions as expected. Interestingly, at energies where most initial ensembles were chaotic, the variation of collisional energy transfer with initial conditions was even more pronounced than in the quasiperiodic regime. Moreover, no discontinuous change in energy transfer moments was apparent as the threshold for chaotic behavior was crossed. This suggests that mode specific effects may be important even when trajectory motion is chaotic, and if this is the case, then the characterization of semiclassical eigenstates in this region is important. Obviously, these conclusions require further study.

The question of the accurate and efficient determination of vibrational action-angle variables is the subject of much of the body of this chapter. A number of methods have now been developed for accurately determining these variables (or the related semiclassical eigenvalues) [3.18,25,29], but few of these methods are efficient enough to permit their use in collisional applications. Certainly, a method cannot use substantially more time for the action-angle analysis than for the trajectory integration if it is to be useful in such applications. To date, the classical perturbation theory approach [3.17], truncated in low order, has proved the most useful in this regard and we describe it in detail in Sect.3.2. This method does have important limitations in many situations, however, since convergence of the expansion is often poor and occasionally nonexistent. Moreover, it has not yet been applied to molecules with more than three atoms, and even for triatomics, it has not been used for the coupled vibrating rotor problem (only for the coupling of vibrational modes after rotation has been approximately separated). Within these limitations, however, it has proven useful in applications to state resolved collisional energy transfer and reaction dynamics as described earlier. Obviously there is still a significant need for improved methods for determining action-angle variables in polyatomic systems, since potential surfaces for collisions involving molecules larger than triatomics are starting to become available, and even for triatomics there are many states [such as the Fermi coupled (100) and (020) states of CO_2] which are not easily described using perturbation theory.

Let us now summarize the contents of the rest of this chapter. We begin in Sect.3.2 with a fairly complete description of the coordinates and coordinate transformations used in studying polyatomic molecule vibrational motions. Also included are sections on the partitioning of vibrational and rotational energies and on the determination of good action-angle variables and semiclassical eigenvalues. In Sect.3.3, the study of polyatomic molecule nonreactive collision processes using the quasiclassical trajectory method is discussed and flowcharts outlining the determination of trajectory initial and final conditions using good action-angle variables are presented. In Sect.3.4 results of an application of the trajectory methods of Sects.3.2,3 to collisional excitation in He + SO_2 are described. This application is given primarily to illustrate the kinds of results which are available from trajectory simulations of collisional energy transfer processes and is not intended for use in modeling experimental relaxation results. Nevertheless, the results of this application do show the potential utility of trajectory methods in describing experimentally relevant collisional energy transfer processes in polyatomic systems.

3.2 Semiclassical Vibration-Rotation Motions of Polyatomic Molecules

We begin our development by presenting the classical mechanical description of polyatomic molecule vibration-rotation motions and semiclassical eigenstates. Most of this discussion will be appropriate for polyatomic molecules with an arbitrary number of atoms N.

3.2.1 Classical Molecular Hamiltonian

Let us first summarize the well-known [3.30] expressions for the molecular internal Hamiltonian in normal coordinates and the transformation from space fixed to molecule fixed variables. This transformation is important in trajectory calculations because the equations of motion are most conveniently integrated in space fixed cartesian coordinates while the internal states of molecules are best defined using normal (or local-mode) coordinates for vibration and Euler angles for rotation.

We assume that the molecule consists of N atoms, each of which is located relative to the origin of an arbitrary space fixed coordinate system (xyz) by a vector $\underline{r}_i$ ($i = 1,\dots,N$). The cartesian components of $\underline{r}_i$ will be denoted $r_{i\alpha}$ ($\alpha = x,y,z$). As discussed elsewhere [3.31], the center of mass motion of the molecule can be easily removed by using cluster coordinates rather than atomic coordinates (in which case there are only N-1 independent $\underline{r}_i$'s) but the development which follows is largely the same using either atomic or cluster coordinates, so atomic coordinates will be used.

The classical expression for the molecular energy is

$$H = \frac{1}{2} \sum_{i=1} m_i \dot{\underline{r}}_i^2 + V \quad , \tag{3.1}$$

where m_i is the mass of atom i and V is the intramolecular potential. The latter depends on only 3N-6 internal variables which will be chosen to be the normal coordinates X_k (k = 1,...3N-6) in this discussion.

In order to interrelate the space fixed cartesian positions $r_{i\alpha}$ of each atom to the normal coordinates, it is necessary to define the displacements of the atomic positions from their equilibrium values. Denoting $\underline{a}_i$ as the equilibrium value of $\underline{r}_i$, the locations of the $\underline{a}_i$'s relative to the $\underline{r}_i$'s are customarily determined by imposing the Eckart condition [3.30]:

$$\sum_i m_i(\underline{a}_i \times \underline{r}_i) = 0 \quad . \tag{3.2}$$

Choosing the molecule fixed coordinate system XYZ to be stationary relative to the molecule at equilibrium, the three components of (3.2) can be used to determine the three Euler angles α, β, γ which interrelate XYZ and xyz. The orientation of XYZ relative to the equilibrium positions of the molecule is, of course, arbitrary, but can often usefully be taken so as to give a diagonal equilibrium moment of inertia tensor (i.e., the principal axis frame for the equilibrium configuration).

With the $\underline{a}_i$'s and hence the Euler angles determined, one can then define displacements from equilibrium via

$$\Delta r_{i\alpha} = r_{i\alpha} - a_{i\alpha} \qquad \begin{array}{l} i = 1,\ldots N \\ \alpha = x,y,z \end{array} \tag{3.3}$$

and in terms of these displacements, the normal coordinates are given by

$$X_k = \sum_i m_i^{\frac{1}{2}} \ell_{ik}^{\alpha} \Delta r_{i\alpha} \qquad k = 1,\ldots 3N-6 \quad . \tag{3.4}$$

Here ℓ_{ik}^{α} is the $i\alpha$'th element of the k'th eigenvector of the vibrational secular equation

$$|\underline{\underline{M}} - \omega_k^2 \, \underline{\underline{I}}| = 0 \tag{3.5}$$

where

$$M_{i\alpha,\, j\beta} = \left. \frac{\partial^2 V}{\partial r_{i\alpha} \partial r_{j\beta}} \right|_{\substack{r_{i\alpha}=a_{i\alpha} \\ r_{j\beta}=a_{j\beta}}} (m_i m_j)^{-\frac{1}{2}} \tag{3.6}$$

$$I_{i\alpha,j\beta} = \delta_{ij}\delta_{\alpha\beta} \tag{3.7}$$

and ω_k is the kth normal mode frequency. Note that if one chooses to set up (3.5) using atomic cartesian coordinates, 6 of the ω_k's will be zero (corresponding to 3 translations and 3 rotations). Often it is convenient to use internal coordinates to set up and solve the secular equation. If that is done, the cartesian eigenvectors ℓ_{ik}^{α} must be constructed using methods which are described elsewhere [3.30b].

The matrix of eigenvectors is orthogonal, so the inverse of (3.4) is

$$\Delta r_{i\alpha} = \sum_{k=1}^{3N-6} \frac{\ell_{ik}^{\alpha}}{\sqrt{m_i}} X_k \quad . \tag{3.8}$$

To define normal momenta P_k conjugate to the X_k's, one first needs to define the molecule fixed velocity components $\partial \underline{r}_i/\partial t$. These are related to the space fixed velocities $\dot{\underline{r}}_i$ via

$$\dot{\underline{r}}_i = \underline{\omega} \times \underline{r}_i + \frac{\partial \underline{r}_i}{\partial t} \tag{3.9}$$

where $\underline{\omega}$ is the angular velocity. $\underline{\omega}$ is determined by calculating the quantity

$$\underline{j}_o = \sum_{i=1}^{N} \underline{a}_i \times \underline{p}_i = \sum_i m_i \underline{a}_i \times (\underline{\omega} \times \underline{r}_i) + \sum_i m_i \left(\underline{a}_i \times \frac{\partial \underline{r}_i}{\partial t} \right) \quad , \tag{3.10}$$

where $\underline{p}_i = m_i \dot{\underline{r}}_i$.

The last term in this expression is zero because of (3.2). Defining the matrix $\underline{\underline{I}}_o$ via

$$(\underline{\underline{I}}_o)_{\alpha\beta} = \sum_i m_i (\underline{a}_i \cdot \underline{r}_i \delta_{\alpha\beta} - a_{i\alpha} r_{i\beta}) \quad , \tag{3.11}$$

one can rearrange (3.10) to

$$\underline{\omega} = \underline{\underline{I}}_o^{-1} \underline{j}_o \tag{3.12}$$

which determines $\underline{\omega}$.

Once (3.9) is used to obtain the body fixed velocity components, the time derivative of (3.4) can be used to determine normal coordinate time derivatives. The normal momenta then follow using [3.30]

$$P_k = \dot{X}_k + \underline{\omega} \cdot \sum_{k'} \underline{\zeta}_{k'k} Q_{k'} \quad , \tag{3.13}$$

where $\underline{\zeta}_{k'k}$ is the Coriolis coupling vector

$$\underline{\zeta}_{k'k} = \sum_i \underline{\ell}_{ik'} \times \underline{\ell}_{ik} \quad . \tag{3.14}$$

In (3.14), the eigenvector elements ℓ_{ik}^{α} are grouped as vectors with components labelled by α.

With these definitions of normal coordinate and momentum variables, the Hamiltonian (3.1) can be expressed as follows [3.30]:

$$H = \frac{1}{2}\sum_k P_k^2 + V(\{X_k\}) + \frac{1}{2}(\underline{j} - \underline{j}_{vib})(\underline{\underline{I}} - \underline{\underline{I}}_{vib})^{-1}(\underline{j} - \underline{j}_{vib}) \quad , \tag{3.15}$$

where $\underline{j}$ is the rotational angular momentum

$$\underline{j} = \sum_i m_i \underline{r}_i \times \dot{\underline{r}}_i \tag{3.16}$$

and $\underline{j}_{vib}$ the vibrational angular momentum

$$\underline{j}_{vib} = \sum_{kk'} \underline{\zeta}_{k'k} X_{k'} P_k \quad . \tag{3.17}$$

$\underline{\underline{I}}$ is the moment of inertia tensor

$$(\underline{\underline{I}})_{\alpha\beta} = \sum_i m_i (r_i^2 \delta_{\alpha\beta} - r_{i\alpha} r_{j\beta}) \quad , \tag{3.18}$$

and $\underline{\underline{I}}_{vib}$ is the vibrational moment of inertia

$$(\underline{\underline{I}}_{vib})_{\alpha\beta} = \sum_{kk'k''} \zeta^{\alpha}_{k'k} \zeta^{\beta}_{k''k} X_{k'} X_k \quad . \tag{3.19}$$

Equation (3.15) is the molecular Hamiltonian which can be used to define semiclassical eigenstates. Because $\underline{j}_{vib}$, $\underline{\underline{I}}$ and $\underline{\underline{I}}_{vib}$ depend on the normal coordinates and momenta, the vibrational and rotational degrees of freedom in H are coupled together. This seriously complicates the determination of molecular action-angle variables as are needed to define semiclassical eigenstates. To date, no fully coupled treatment of this problem has been developed for molecules other than diatomics. Approximate solutions can be obtained, however, which we now discuss.

3.2.2 Rotational-Vibrational Partitioning

Perhaps the most obvious approach to the simplification of the Hamiltonian in (3.15) for the purpose of determining semiclassical eigenvalues is to introduce approximate separations between vibration and rotation. The validity of these separations depends rather strongly on the differences in timescale between vibration and rotation. Such differences often exist for the low vibrational and rotational states of molecules, but not for highly rotationally excited molecules or for "floppy" vibrational motions.

In two rather special limits, the Hamiltonian in (3.15) can be exactly separated between rotation and vibration. One occurs when the total molecular angular momentum $\underline{j}$ equals zero. In that case, a purely vibrational Hamiltonian

$$H_{vib} = \frac{1}{2}\sum_k P_k^2 + V(\{X_k\}) + \frac{1}{2}\underline{j}_{vib}(\underline{\underline{I}} - \underline{\underline{I}}_{vib})^{-1}\underline{j}_{vib} \tag{3.20}$$

is obtained. The last term in this expression represents the zero $\underset{\sim}{j}$ Coriolis contribution to the internal energy and is usually very small for molecules in their ground and lowest excited states. The determination of semiclassical eigenstates for vibrational motion described by H_{vib} is described in Sect.3.2.2.

The second limit where H simplifies considerably occurs in the limit of zero vibrational energy. In this case, we obtain a purely rotational (rigid rotor) Hamiltonian

$$H_{rot} = \frac{1}{2}\, \underset{\sim}{j}\,\underset{\approx}{I}_e^{-1}\,\underset{\sim}{j} \tag{3.21}$$

[where $\underset{\approx}{I}_e$ is defined by (3.18), replacing $r_{i\alpha}$ by $a_{i\alpha}$ everywhere]. Semiclassical eigenstates for this system are discussed in Sect.3.2.4.

In the limit where both the vibrational and rotational excitation are small, a logical partitioning of the full Hamiltonian in (3.19) is

$$H_1 = H_{vib} + H_{rot} \quad , \tag{3.22}$$

where (3.20,21) are used for H_{vib} and H_{rot}. In this expression, vibration and rotation are completely separated, with no dependence of vibrational state on $\underset{\sim}{j}$ and no dependence of the rotational state on the vibrational quantum numbers. The validity of this approximation depends on the magnitude of the difference H_C between H and H_1. H_C, which is simply the nonzero $\underset{\sim}{j}$ part of the Coriolis Hamiltonian, depends on the relative size of $\underset{\sim}{j}$ to the vibrational angular momentum $\underset{\sim}{j}_{vib}$. Since the latter quantity is of the order of $\hbar$ for molecules in their vibrational ground state, we find that $j \gg \hbar$ is needed to make $H_{rot} \gg H_C$. At the same time, one cannot use $\underset{\sim}{j}$'s so large that $H_{rot} \gg H_{vib}$, for in that case, H_C may be comparable to H_{vib}.

For small $\underset{\sim}{j}$'s, one often finds that $H_{vib} \gg H_{rot}$ and $H_{vib} \gg H_C$ but that H_{rot} and H_C are comparable. In that case, an alternative partitioning is

$$H = H_{vib} + H'_{rot} \quad , \tag{3.23}$$

where $H'_{rot} = H_{rot} + H_C$. In Sect.3.3, a number of trajectory results are presented using this partitioning. This partitioning leads to rotational states which depend on vibrational energy but the vibrational states are still $\underset{\sim}{j}$ independent.

For situations where H_{vib} and H_{rot} are comparable, the above partitionings are inappropriate. If H_C is still small compared to H_{vib} and H_{rot}, a self-consistent treatment is possible wherein the rotational Hamiltonian is taken as the average of H over vibrational motions, while the vibrational Hamiltonian is an average of H over rotation. Alternatively, the Hamiltonian can be partitioned using

$$H = H'_{vib} + H_{rot} \quad , \tag{3.24}$$

where $H'_{vib} = H_{vib} + H_C$. This would be appropriate when H_{vib} and H_{rot} are comparable but there is still an adequate timescale separation between vibration and rotation to make the body fixed rotational angular momentum components nearly constant on the vibrational timescale. In this case, H_C serves as an effective centrifugal potential governing vibrational motions. While this approach is known to work well for diatomic molecule systems [3.32], very little experience concerning its application to polyatomics exists. Obviously the preferred procedure is to avoid partitioning H at all and solve the fully coupled vibration-rotation problem. The practical difficulties associated with doing this at present make the development of accurate partitionings such as are presented here important.

3.2.3 Vibrational Semiclassical Eigenstates: Classical Perturbation Theory

In this section we assume that a partitioning such as in (3.21) or (3.23) has been made and that we wish to determine the semiclassical eigenstates associated with the vibrational Hamiltonian in (3.20). For trajectory studies it is also important to determine the explicit transformation between normal coordinate and momentum variables (X_k, P_k) and the good action and angle variables N_k and Q_k.

For the present studies, we consider only the case of quasiperiodic (nonchaotic), vibrational motion [3.25], where 3N-6 good constants of the motion exist which are the actions N_k. Semiclassical eigenstates are defined much as in one-dimensional WKB theory by equating these actions to integer multiples of $\hbar$ (here the convention of [3.18] is used wherein the zero point term is incorporated into the action definition). The transformation between normal coordinate variables and action-angle variables is conveniently accomplished using a generating function F_2 which is obtained by solving the molecular Hamilton-Jacobi equation. Since this is a 3N-6 dimensional *nonlinear* partial differential equation, it is in some respects more difficult to solve than the molecular Schrödinger equation. However, even during the days of the Old Quantum Theory [3.24], it was known that *classical perturbation theory* was a simple and efficient approach to its solution, and more recent numerical studies [3.18] (see below) have indicated its surprising accuracy as well. For trajectory applications it is currently the only method with sufficient efficiency to make the initial and final semiclassical eigenstate determination significantly less time consuming than the trajectory integration [3.17].

To develop perturbation theory one ordinarily chooses some reference Hamiltonian for which action-angle variables are analytically known. An obvious choice is the separable normal mode Hamiltonian H_{SNM}. Denoting the separable normal mode action-angle variables as n_k and q_k (k = 1,...3N-6), then H_{SNM} is given by

$$H_{SNM} = \frac{1}{2} \sum_k (P_k^2 + \omega_k^2 X_k) = \sum_k \omega_k (n_k + \tfrac{1}{2}) \quad , \tag{3.25}$$

where we take $\hbar = 1$ throughout this discussion. With these definitions we can write

$$H_{vib} = H_{SNM} + H'(\{X_k, P_k\}) \quad , \tag{3.26}$$

where H' combines all the anharmonic terms in V (3.20) with the Coriolis term. Note that H' depends on both the normal mode coordinates *and* momenta, and that these are in turn dependent on n_k and q_k through

$$X_k = \left(\frac{2(n_k + \frac{1}{2})}{\omega_k}\right)^{\frac{1}{2}} \cos q_k \tag{3.27a}$$

$$P_k = -[2\omega_k(n_k + \tfrac{1}{2})]^{\frac{1}{2}} \sin q_k \quad . \tag{3.27b}$$

We now wish to consider the canonical transformation between n_k, q_k and N_k, Q_k. This is specified in terms of F_2 via

$$n_k = \frac{\partial F_2(N_k, q_k)}{\partial q_k} \tag{3.28a}$$

$$Q_k = \frac{\partial F_2(N_k, q_k)}{\partial N_k} \quad . \tag{3.28b}$$

Substitution of (3.28a) into (3.27a) and (3.27b), and the latter into (3.26), using (3.25) leads to the Hamilton-Jacobi equation for F_2:

$$\sum_k \omega_k\left(\frac{\partial F_2}{\partial q_k} + \frac{1}{2}\right) + H'\left[X_k\left(\frac{\partial F_2}{\partial q_k}, q_k\right), P_k\left(\frac{\partial F_2}{\partial q_k}, q_k\right)\right] = E_{vib} \tag{3.29}$$

where E_{vib} is the total vibrational energy [a constant of the motion for the separated Hamiltonian (3.20)].

To solve (3.21) for F_2, we use the fact [3.18] that the difference between F_2 and the identity transformation $\underline{q} \cdot \underline{N}$ is periodic in $\underline{q}$. Expanding this difference in a Fourier series, we find

$$F_2(N_k, q_k) = \underline{q} \cdot \underline{N} - i \sum_{\underline{\ell}} A_{\underline{\ell}}\, e^{i\underline{\ell} \cdot \underline{q}} \tag{3.30}$$

with $\underline{\ell}$ a vector containing 3N-6 integers, each varying from $-\infty$ to ∞.

Substitution of (3.30) into the Hamilton-Jacobi equation yields

$$E_{vib} = \underline{\omega} \cdot (\underline{N} + \frac{1}{2}) + \sum_{\underline{\ell}} (\underline{\omega} \cdot \underline{\ell}) A_{\underline{\ell}}\, e^{i\underline{\ell} \cdot \underline{q}} + H' \quad . \tag{3.31}$$

Taking Fourier components of (3.31) we find

$$A_{\underline{\ell}} = \frac{-1}{\underline{\omega} \cdot \underline{\ell}} \int_0^{2\pi} \frac{dq_1}{2\pi} \int_0^{2\pi} \frac{dq_2}{2\pi} \cdots \int_0^{2\pi} \frac{dq_{3N-6}}{2\pi} H'\, e^{-i\underline{\ell} \cdot \underline{q}} \tag{3.32}$$

for $\underline{\ell} \neq 0$, and

$$E_{vib} = \underline{\omega} \cdot (\underline{N} + \frac{1}{2}) + \int_0^{2\pi} \frac{dq_1}{2\pi} \cdots \int_0^{2\pi} \frac{dq_{3N-6}}{2\pi} H' \tag{3.33}$$

for $\underline{\ell} = 0$.

The essence of the perturbation approach is to assume that in zeroth order, all of the $A_{\underline{\ell}}^{(0)}$'s are zero and $E_{vib}^{(0)} = \underline{\omega} \cdot (\underline{N} + \frac{1}{2})$. In first order, the $A_{\underline{\ell}}^{(0)}$'s are used in (3.33) to define $E_{vib}^{(1)}$ and in (3.27,28a,30,32) to define $A_{\underline{\ell}}^{(1)}$'s (expanding square roots where necessary to first order). In second order, the $A_{\underline{\ell}}^{(1)}$'s are substituted into (3.33) to give $E_{vib}^{(2)}$ and into (3.27,28a,30,32) to determine $A_{\underline{\ell}}^{(2)}$'s. Successive orders proceed analogously with this iterative process. There is, of course, no guarantee that the process will converge, and in fact divergence does occur in the chaotic regime. Nevertheless, for the lowest energy levels of molecules, convergence is often rapid.

To illustrate this application, let us consider a nonlinear triatomic molecule with a quartic force field potential. This example is of significant practical importance because quartic (but not higher) force fields have been determined spectroscopically for many triatomics. We will suppose for simplicity that the triatomic has two identical end atoms (i.e., AB_2), in which case the quartic force field may be expressed as

$$V = \frac{1}{2}\,(\omega_1^2 X_1^2 + \omega_2^2 X_2^2 + \omega_3^2 X_3^2) + V_C + V_Q \quad , \tag{3.34}$$

where V_C and V_Q are the cubic and quartic anharmonic terms:

$$V_C = \lambda_1 X_1^3 + \lambda_2 X_2^3 + \lambda_{12} X_1 X_2^2 + \lambda_{21} X_1^2 X_2 + \lambda_{13} X_1 X_3^2 + \lambda_{23} X_2 X_3^2 \tag{3.35}$$

$$V_Q = \gamma_1 X_1^4 + \gamma_2 X_2^4 + \gamma_3 X_3^4 + \gamma_{12} X_1^2 X_2^2 + \gamma_{13} X_1^2 X_3^2 + \gamma_{23} X_2^2 X_3^2 \quad . \tag{3.36}$$

Note that the modes are ordered such that the two totally symmetric modes (symmetric stretch and bend) are labelled 1 and 2 respectively and the antisymmetric (or asymmetric) stretch is mode 3. All of the nonzero cubic terms have been included in (3.35) but (3.36) ignores certain nonzero quartic terms (such as $X_1^3 X_2$) which do not contribute to the semiclassical eigenvalues in low order.

Before proceeding with the perturbation theory evaluation, it should be noted that V_C does not contribute to the semiclassical eigenvalues in first order and its second-order contribution is (typically) of the same order in magnitude as the first-order contribution of V_Q. Because of this we will carry only V_C through the first iteration of the perturbation cycle described above, then pick up V_Q in the second iteration (wherein V_C is treated in second order and V_Q to first).

The Coriolis term in (3.20) is also included in the perturbation theory evaluation. Now since the normal mode eigenvectors $\underline{\ell}_{ik}$ are necessarily restricted to be in the plane formed by the 3 atoms, (3.14) indicates that $\underline{\zeta}_{k'k}$ must be perpendicular to this plane (a direction which we denote as the Y direction). In addition, the C_{2v} symmetry of the triatomic leads to $\underline{\zeta}_{12} = \underline{\zeta}_{21} = 0$ so that the only nonzero $\underline{\zeta}_{k'k}$ components are $\zeta_{13}^Y = -\zeta_{31}^Y$ and $\zeta_{23}^Y = -\zeta_{32}^Y$. Using this in (3.17) we find that only the Y component of $\underline{j}_{vib}$ is nonzero:

$$(\underline{j}_{vib})_Y = \zeta_{13}^Y(X_1P_3 - X_3P_1) + \zeta_{23}^Y(X_2P_3 - X_3P_2) \quad . \tag{3.37}$$

Using this result for $\underline{j}_{vib}$ in (3.20) yields the Coriolis Hamiltonian

$$H_C = \frac{1}{2}(\underline{\underline{I}} - \underline{\underline{I}}_{vib})^{-1}_{YY}(\underline{j}_{vib})^2_Y \quad . \tag{3.38}$$

Both $\underline{\underline{I}}$ and $\underline{\underline{I}}_{vib}$ depend on the normal coordinates, but $\underline{\underline{I}}$ also contains a larger coordinate independent part which represents the moment of inertia of the triatomic at equilibrium. Since the contribution of H_C to the total energy is much smaller than that of V_C or V_Q, we are justified in neglecting the weak dependence of $(\underline{\underline{I}} - \underline{\underline{I}}_{vib})^{-1}$ on normal coordinates, replacing it by $(\underline{\underline{I}}_e)^{-1}$, see (3.21). H_C thus becomes

$$H_C = \frac{1}{2} I^{-1}_{eYY}[\zeta_{13}^Y(X_1P_3 - X_3P_1) - \zeta_{23}^Y(X_2P_3 - X_3P_2)]^2 \quad . \tag{3.39}$$

If (3.27) are substituted into (3.39), the resulting expression for H_C will depend on products of the action variables to an overall power of two. This is the same order of dependence as is obtained from V_Q which suggests that H_C should be grouped with V_Q in performing the perturbation iteration.

Let us now begin the perturbation theory evaluation. In zeroth order, the Fourier coefficients $A_\ell^{(0)}$ are all zero, while the zeroth order energy is

$$E_{vib}^{(0)} = \sum_{i=1}^{3} \omega_i J_i \quad . \tag{3.40}$$

Here we have introduced the abbreviation $J_i = N_i + \frac{1}{2}$ (i = 1,2,3). Considering V_C in first order, we find that the first-order energy $E_{vib}^{(1)}$ is identical to $E_{vib}^{(0)}$, while the following 12 $A_\ell^{(1)}$'s are nonzero:

$$A_{100}^{(1)} = -\left[\frac{3\lambda_1}{8}\left(\frac{2J_1}{\omega_1}\right)^{3/2} + \frac{\lambda_{12}}{4}\left(\frac{2J_1}{\omega_1}\right)^{1/2}\left(\frac{2J_2}{\omega_2}\right) + \frac{\lambda_{13}}{4}\left(\frac{2J_1}{\omega_1}\right)^{1/2}\left(\frac{2J_3}{\omega_3}\right)\right]/\omega_1 \tag{3.41a}$$

$$A_{300}^{(1)} = \frac{-\lambda_1}{24\omega_1}\left(\frac{2J_1}{\omega_1}\right)^{3/2} \tag{3.41b}$$

$$A_{010}^{(1)} = -\left[\frac{3\lambda_2}{8}\left(\frac{2J_2}{\omega_2}\right)^{3/2} + \frac{\lambda_{21}}{4}\left(\frac{2J_1}{\omega_1}\right)\left(\frac{2J_2}{\omega_2}\right)^{1/2} + \frac{\lambda_{23}}{4}\left(\frac{2J_2}{\omega_2}\right)^{1/2}\left(\frac{2J_3}{\omega_3}\right)\right]/\omega_2 \tag{3.41c}$$

$$A_{030}^{(1)} = \frac{-\lambda_2}{24\omega_2}\left(\frac{2J_2}{\omega_2}\right)^{3/2} \tag{3.41d}$$

$$A_{120}^{(1)} = \frac{-\lambda_{12}}{8(\omega_1 + 2\omega_2)}\left(\frac{2J_1}{\omega_1}\right)^{1/2}\left(\frac{2J_2}{\omega_2}\right) \tag{3.41e}$$

$$A^{(1)}_{1-20} = \frac{-\lambda_{12}}{8(\omega_1 - 2\omega_2)}\left(\frac{2J_1}{\omega_1}\right)^{1/2}\left(\frac{2J_2}{\omega_2}\right) \tag{3.41f}$$

$$A^{(1)}_{210} = \frac{-\lambda_{21}}{8(2\omega_1 + \omega_2)}\left(\frac{2J_1}{\omega_1}\right)\left(\frac{2J_2}{\omega_2}\right)^{1/2} \tag{3.41g}$$

$$A^{(1)}_{2-10} = \frac{-\lambda_{21}}{8(2\omega_1 - \omega_2)}\left(\frac{2J_1}{\omega_1}\right)\left(\frac{2J_2}{\omega_2}\right)^{1/2} \tag{3.41h}$$

$$A^{(1)}_{102} = \frac{-\lambda_{13}}{8(\omega_1 + 2\omega_3)}\left(\frac{2J_1}{\omega_1}\right)^{1/2}\left(\frac{2J_3}{\omega_3}\right) \tag{3.41i}$$

$$A^{(1)}_{10-2} = \frac{-\lambda_{13}}{8(\omega_1 - 2\omega_3)}\left(\frac{2J_1}{\omega_1}\right)^{1/2}\left(\frac{2J_3}{\omega_3}\right) \tag{3.41j}$$

$$A^{(1)}_{012} = \frac{-\lambda_{23}}{8(\omega_2 + 2\omega_3)}\left(\frac{2J_2}{\omega_2}\right)^{1/2}\left(\frac{2J_3}{\omega_3}\right) \tag{3.41k}$$

$$A^{(1)}_{01-2} = \frac{-\lambda_{23}}{8(\omega_2 - 2\omega_3)}\left(\frac{2J_2}{\omega_2}\right)^{1/2}\left(\frac{2J_3}{\omega_3}\right) \quad . \tag{3.41l}$$

It should be noted here that there are many other nonzero $A^{(1)}_{\underline{\ell}}$'s, but all of these are related by simple symmetry relations (such as $A_{-\underline{\ell}} = -A_{\underline{\ell}}$) to the above set. The above set is all that is needed if we use these symmetry relations to rewrite (3.30) as

$$F_2 = \underline{q} \cdot \underline{N} + 2 \sum_{\underline{\ell}} A_{\underline{\ell}} \sin(\underline{\ell} \cdot \underline{q}) \quad . \tag{3.42a}$$

In second order, we use these $A^{(1)}_{\underline{\ell}}$'s in (3.42a), then (3.28a) and (3.27) to express the coordinates and momenta appearing in H' in (3.33). Adding in the first-order energy expressions obtained from $H_Q + H_C$ we obtain the following expression for the second-order eigenvalues:

$$E^{(2)}_{vib} = \sum_{i=1}^{3} \omega_i J_i + \sum_{i=1}^{3} \sum_{j=1}^{3} F_{ij} J_i J_j \tag{3.42b}$$

with

$$F_{11} = \frac{-15\lambda_1^2}{4\omega_1^4} - \frac{\lambda_{21}^2(3\omega_2^2 - 8\omega_1^2)}{4\omega_1^2\omega_2^2(\omega_2^2 - 4\omega_1^2)} + \frac{3\gamma_1}{2\omega_1^2} \tag{3.43a}$$

$$F_{22} = \frac{-15\lambda_2^2}{4\omega_2^4} - \frac{\lambda_{12}^2(3\omega_1^2 - 8\omega_2^2)}{4\omega_1^2\omega_2^2(\omega_1^2 - 4\omega_2^2)} + \frac{3\gamma_2}{2\omega_2^2} \tag{3.43b}$$

$$F_{33} = \frac{-\lambda_{13}^2}{4\omega_1^2\omega_3^2}\left(\frac{3\omega_1^2 - 8\omega_3^2}{\omega_1^2 - 4\omega_3^2}\right) - \frac{\lambda_{23}^2}{4\omega_2^2\omega_3^2}\left(\frac{3\omega_2^2 - 8\omega_3^2}{\omega_2^2 - 4\omega_3^2}\right) + \frac{3\gamma_3}{2\omega_3^2} \tag{3.43c}$$

$$F_{12} = \frac{-3\lambda_1\lambda_{12}}{\omega_1^3\omega_2} - 3\,\frac{\lambda_2\lambda_{21}}{\omega_2^3\omega_1} + \frac{2\lambda_{12}^2}{\omega_1\omega_2(\omega_1^2 - 4\omega_2^2)} \tag{3.43d}$$

$$+ \frac{2\lambda_{21}^2}{\omega_1\omega_2(\omega_2^2 - 4\omega_1^2)} + \frac{\gamma_{12}}{\omega_1\omega_2}$$

$$F_{13} = -\,3\,\frac{\lambda_1\lambda_{13}}{\omega_1^3\omega_3} - \frac{\lambda_{21}\lambda_{23}}{\omega_1\omega_2^2\omega_3} + \frac{2\lambda_{13}^2}{\omega_1\omega_3(\omega_1^2 - 4\omega_3^2)} + \frac{\gamma_{13}}{\omega_1\omega_3} \tag{3.43e}$$

$$+ \frac{1}{2}\,\zeta_{13}^{Y2} I_{eYY}^{-1}(\omega_3/\omega_1 + \omega_1/\omega_3)$$

$$F_{23} = -\,3\,\frac{\lambda_2\lambda_{23}}{\omega_2^3\omega_3} - \frac{\lambda_{12}\lambda_{13}}{\omega_1^2\omega_2\omega_3} + \frac{2\lambda_{23}^2}{\omega_2\omega_3(\omega_2^2 - 4\omega_3^2)} + \frac{\gamma_{23}}{\omega_2\omega_3} \tag{3.43f}$$

$$+ \frac{1}{2}\,\zeta_{23}^{Y2} I_{eYY}^{-1}(\omega_3/\omega_2 + \omega_2/\omega_3) \quad .$$

The second-order $A_\ell^{(2)}$'s are obtained using the first-order expressions for the coordinates and momenta in evaluating H' in (3.32). In doing this, we find that all of the above $A_\ell^{(1)}$'s are unchanged in going to second order. There are, however, 58 additional nonzero $A_\ell^{(2)}$'s. Because the explicit expressions for these are very complex, they will not be given here. Since they are important to the determination of accurate action-angle variables, a program which calculates them is available through the NRCC software catalog [3.33] (or by contacting this author).

Higher-order perturbation theory evaluations are possible using numerical iteration methods [3.18,34] and are often important in obtaining accurate energy levels for the vibrationally excited states of polyatomics (see below). For collisional applications, we have found these additional iterations to be quite time consuming (remember, they have to be applied to each trajectory). In addition, the precision obtained using a 2nd-order evaluation with the analytical expressions described above is often adequate. For this reason, we have used the 2nd-order treatments in the studies presented in Sect.3.3 and in other work [3.19,23].

To conclude this section, we use the zeroth and second-order energy expressions given above (3.40,42), along with a 3rd-order energy obtained numerically by evaluating (3.33) with 2nd-order expressions for the coordinate and momenta substituted, in an evaluation of the semiclassical eigenvalues for several states of SO_2 and H_2O. The results are presented in Table 3.1. The states considered for each molecule

Table 3.1. Semiclassical eigenvalues for nonlinear nonrotating triatomic molecules[a]

A. SO_2[b] N_1 N_2 N_3	Harmonic	2nd Order	3rd Order	Exact Semiclassical[c]	Exact Quantum[d]
0 0 0	1536.82	1528.88	1528.95	1529.12	1529.60
1 0 0	2707.91	2684.15	2684.67	2685.11	2685.63
0 1 0	2061.58	2044.67	2044.57	2045.36	2045.81
0 2 0	2586.35	2554.49	2553.91	2555.86	2556.21
0 0 1	2914.62	2887.61	2888.06	2889.07	2889.53

B. H_2O[c] N_1 N_2 N_3	Harmonic	2nd Order	3rd Order	Exact Semiclassical[c]	Exact Quantum[c]
0 0 0	4710.72 cm^{-1}	4626.95	4629.23	4645.11	4651.98
1 0 0	8543.62	8283.89	8301.41	8360.90	8369.29
0 1 0	6357.48	6208.77	6215.66	6242.55	6249.33
0 2 0	8004.24	7757.07	7753.16	7804.80	7811.53
0 0 1	8652.52	8368.08	8286.57	8466.58	8472.75

[a] All energies are in cm^{-1}; [b] Quartic force field of [3.35]; [c] [3.34]; [d] [3.37]; [e] Quartic force field of [3.36].

are (000), (100), (010), (020), and (001). Listed first is the simple harmonic (zeroth order) semiclassical eigenvalue for each state, then the 2nd and 3rd-order perturbation results, the exact semiclassical eigenvalues of HANDY et al. [3.34] and the corresponding exact quantum eigenvalues of WHITEHEAD and HANDY [3.37].

For SO_2, Table 3.1 indicates that the perturbation theory result converges quite rapidly. The ground state 2nd-order eigenvalue differs by only 0.24 cm^{-1} from the exact semiclassical result and the excited state eigenvalues are, in 2nd order, off by less than 1.5 cm^{-1}. Some of the 3rd-order eigenvalues are better than 2nd, but some are worse. This could be the result of improper balancing of positive and negative anharmonic terms in 3rd order. As was noted in [3.17] for two mode systems, the cubic and quartic corrections to the total energy are opposite in sign but similar in magnitude so the net anharmonic correction involves substantial cancellation. Evidently our 2nd-order treatment (wherein the quartic terms are only treated to first order) accurately weighs the positive and negative contributions, while the 3rd-order treatment overemphasizes the negative contributions due to the cubic term, thus causing less accurate eigenvalues in some cases.

The comparison of exact semiclassical with perturbation theory eigenvalues for H_2O indicates that the convergence is much slower than for SO_2, with the 2nd-order ground state energy off by 18 cm^{-1} and other states in error by as much as

98.5 cm^{-1}. This is not surprising, for the size of the anharmonic correction for H_2O (66 cm^{-1} for the ground state) is much larger than for SO_2 (7.7 cm^{-1}). Again the 3rd-order result is not uniformly better than the 2nd-order results. It is also the case that the exact semiclassical eigenvalue differs from the exact quantum result by more for H_2O (6.87 cm^{-1} for the ground state) than for SO_2 (0.48 cm^{-1}). The differences between exact semiclassical and exact quantum eigenvalues for both H_2O and SO_2 have been discussed in detail in [3.34], where it was noted that this difference is almost independent of quantum state. This causes the energy differences between different eigenvalues to be more accurate than the eigenvalues themselves. Unfortunately, the perturbation results do not share this characteristic; the eigenvalues for the lower states tend to be more accurate.

3.2.4 Rotational Motions and Eigenstates

As discussed in Sect.3.2.2, the precise form of the rotational Hamiltonian depends significantly on how vibration and rotation are partitioned. The simplest partitioning was that of (3.22) which used H_{rot} from (3.21) (i.e., the rigid molecule Hamiltonian). In what follows we shall assume this has been used. The generalization of the development to the fully coupled Hamiltonian (3.15) will be discussed at the end of this section.

There are many different variables which can be used to describe rotational motion. Perhaps the most obvious set are the Euler angles ϕ, θ and ψ which interrelate the space-fixed coordinates x, y, z to the Eckart frame coordinates X, Y, Z. The angles may be determined at any time via (3.2).

The expression for H_{rot} in terms of ϕ, θ and ψ has been derived by AUGUSTIN and MILLER [3.38] and is

$$H_{rot} = \frac{1}{2}\,[I_X^e(\sin\psi\dot{\theta} - \cos\psi\sin\theta\dot{\phi})^2 + I_Y^e(\cos\psi\dot{\theta} + \sin\psi\sin\theta\dot{\phi})^2 + I_Z^e(\dot{\psi} + \cos\theta\dot{\phi})^2] \quad . \tag{3.44}$$

In this expression, we have assumed that the Eckart frame coordinate system X Y Z coincides with the principal axis frame at equilibrium so that I_X^e, I_Y^e and I_Z^e are the principal moments of inertia. The coefficients of I_X^e, I_Y^e and I_Z^e in (3.44) are, of course, the angular velocities ω_X, ω_Y and ω_Z associated with rotation of the molecule fixed coordinate system. If (3.44) is re-expressed in terms of the momenta P_ϕ, P_θ, and P_ψ (where $P_\phi = \partial H_{rot}/\partial\phi$, $P_\theta = \partial H_{rot}/\partial\theta$, $P_\psi = \partial H_{rot}/\partial\psi$), we get

$$H_{rot} = \frac{1}{2I_X^e}\left(-\frac{P_\phi\cos\psi}{\sin\theta} + P_\theta\sin\psi + P_\psi\cot\theta\cos\psi\right)^2$$

$$+ \frac{1}{2I_Y^e}\left(\frac{P_\phi \sin\psi}{\sin\theta} + P_\theta\cos\psi - P_\psi\cot\theta\sin\psi\right)^2$$

$$+ \frac{1}{2I_Z^e} P_\psi^2 \quad . \tag{3.45}$$

Hamilton's equations for rigid body rotation using the variables ϕ, θ, ψ, P_ϕ, P_θ, P_ψ are easily determined from (3.45).

If we wish to define semiclassical eigenvalues using H_{rot}, it is convenient to reexpress H_{rot} in terms of the symmetric top rotational action variables j, m, k and their conjugate angles χ_j, χ_m and χ_k. These actions are the three good constants of the motion for a symmetric top, namely, the magnitude of the angular momentum (j), the projection of $\underline{j}$ along z (m) and the projection of $\underline{j}$ along Z (k). In terms of the Euler angles and momenta, j, m and k are given by:

$$j = \left[P_\theta^2 + (P_\phi^2 + P_\psi^2 - 2P_\phi P_\psi \cos\theta)/\sin^2\theta\right]^{\frac{1}{2}}$$

$$= (j_X^2 + j_Y^2 + j_Z^2)^{\frac{1}{2}} \tag{3.46a}$$

$$m = P_\phi = j_z \tag{3.46b}$$

$$k = P_\psi = j_Z \quad . \tag{3.46c}$$

The conjugate angles are given by:

$$\chi_j = \cos^{-1}\left[\frac{j^2\cos\theta - mk}{(j^2 - m^2)^{\frac{1}{2}}(j^2 - k^2)^{\frac{1}{2}}}\right] \tag{3.47a}$$

$$\chi_m = \phi - \cos^{-1}\left[\frac{m\cos\theta - k}{\sin\theta(j^2 - m^2)^{\frac{1}{2}}}\right] \tag{3.47b}$$

$$\chi_k = \psi - \cos^{-1}\left[\frac{k\cos\theta - m}{\sin\theta(j^2 - k^2)^{\frac{1}{2}}}\right] \quad . \tag{3.47c}$$

Note that the inverse cosines in (3.47) should be replaced by their supplement if $P_\theta < 0$.

Using (3.46,47) in (3.45), the following expression for H_{rot} is obtained:

$$H_{rot} = j^2\left(\frac{\cos^2\chi_k}{2I_X^e} + \frac{\sin^2\chi_k}{2I_Y^e}\right) + k^2\left(\frac{1}{2I_Z^e} - \frac{\cos^2\chi_k}{2I_X^e} - \frac{\sin^2\chi_k}{2I_Y^e}\right) \quad . \tag{3.48}$$

Equation (3.48) is a very convenient form for the rotational Hamiltonian since it incorporates the two rigorous constants of the motion j and m. k is a constant of the motion if $I_X^e = I_Y^e$ (i.e., for a symmetric top) because H_{rot} is independent of χ_k.

In the more general situation of an asymmetric top, k is not a constant of the motion but H_{rot} is [assuming the validity of (3.22)]. The good action K describing asymmetric top motion can be obtained from the usual integral

$$K = \frac{1}{2\pi}\int_0^{2\pi} k(\chi_k)d\chi_k \tag{3.49}$$

with the χ_k dependence of k obtained from (3.48). Semiclassical eigenvalues are then obtained by applying the usual WKB prescription to K. This is complicated somewhat by the fact that k (χ_k) can exhibit anywhere from 0 to 4 turning points (spanning the limits from free rotor to hindered rotor motion). COLWELL et al. [3.39] have studied this point in some detail with excellent comparisons of semiclassical and quantum eigenvalues in applications to rigid polyatomic molecule rotational states.

Let us now discuss the perturbation solution to the full vibrating rotor problem. Equation (3.15) is the starting point, with P_k and X_k replaced by the harmonic action-angle variables n_k and q_k and $\underline{j}$ expressed in terms of the rotational action-angle variables using

$$j_X = \cos\chi_k(j^2 - k^2)^{\frac{1}{2}} \tag{3.50a}$$

$$j_Y = \sin\chi_k(j^2 - k^2)^{\frac{1}{2}} \tag{3.50b}$$

$$j_Z = k \quad . \tag{3.50c}$$

One then expands the generating function analogous to F_2 in (3.30) in a Fourier expansion in the variables q_k $(k = 1,\ldots,3N\text{-}6)$ and χ_k. By taking Fourier components of (3.30) with respect to q_k and χ_k, a set of *nonlinear* equations for the Fourier coefficients is obtained which can be solved by iteration. To date, no applications of this type have been reported. This is because (a) the simplest realistic example to which applications might be attempted, namely a nonlinear triatomic molecule, requires a 4-dimensional Fourier expansion (the three normal modes plus χ_k) involving hundreds of coefficients, and (b) the use of symmetric top action-angle variables as a basis for Fourier expanding the generating function of an asymmetric top molecule is highly suspect for many triatomic molecules. While it is certainly possible to improve the choice of zeroth-order rotational action-angle variables, the difficulties associated with solving nonlinear equations involving hundreds of Fourier coefficients remains unresolved at this point.

3.3 Trajectory Studies of Polyatomic Molecule Collisions

In this section we describe how the methods developed in the previous section for describing the bound states of polyatomic molecules can be used in trajectory studies of state resolved polyatomic molecule collision processes, particularly collisional energy transfer. This section consists of three parts. In Sects.3.3.1,2, the determination of trajectory initial and final conditions is discussed, while in Sect.3.3.3, the assignment of final state quantum numbers is considered.

3.3.1 Trajectory Initial Conditions

Here we consider the evaluation of trajectory initial conditions (coordinates and momenta) for a polyatomic molecule which is initially in a semiclassical eigenstate. These initial conditions could, of course, be specified in any coordinate system, but general experience with trajectory calculations [3.40] has lead to the widespread use of space-fixed cartesian coordinates for this purpose. The semiclassical eigenstates are, however, best defined using good action-angle variables (choosing the actions according to some quantization scheme and varying the conjugate angles randomly over a 2π interval). Thus we are faced with determining the transformation between action-angle variables and cartesian variables. For the purpose of this discussion, we will assume that the simple partitioning between vibration and rotation given by (3.22) is adequate. This allows us to define vibrational and rotational eigenstates separately, although this information must obviously be combined together in determining cartesian variables. We also assume that the perturbation procedure outlined in Sect.3.2.3 is adequate to define the vibrational good action-angle variables and that the symmetric top rotational action-angle variables are adequate to define rotational states.

A schematic diagram of the initial condition determination process is given in Table 3.2. Starting with the vibrational good action-angle variables, the most numerically time-consuming step in this process is the first one in which (3.28) are solved to determine the harmonic variables n_k and q_k. Equation 3.28a is a simple summation, but (3.28b) requires the solution of nonlinear equations for q_i given N_i and Q_i. After that, (3.27) are used to determine the normal mode coordinates and momenta. The normal mode coordinates X_i are then transformed via (3.8) to cartesian coordinates and the latter are then rotated via the Euler angles from the body-fixed to the space-fixed frames. The normal momenta P_i are converted to velocities using (3.13) and the time derivatives relative to the rotating molecular frame are converted to derivatives relative to the space-fixed frame using (3.9). Finally the velocity components $dr_{i\alpha}/dt$ are transformed to the space-fixed frame using an Euler angle rotation.

The Euler angles ϕ, θ, ψ are obtained from j, m, k, χ_j, χ_m and χ_k using the inverse of (3.47). The angular velocities ω_X, ω_Y and ω_Z, which are used in the velo-

Table 3.2. Schematic diagram of initial condition determination process for polyatomic molecules

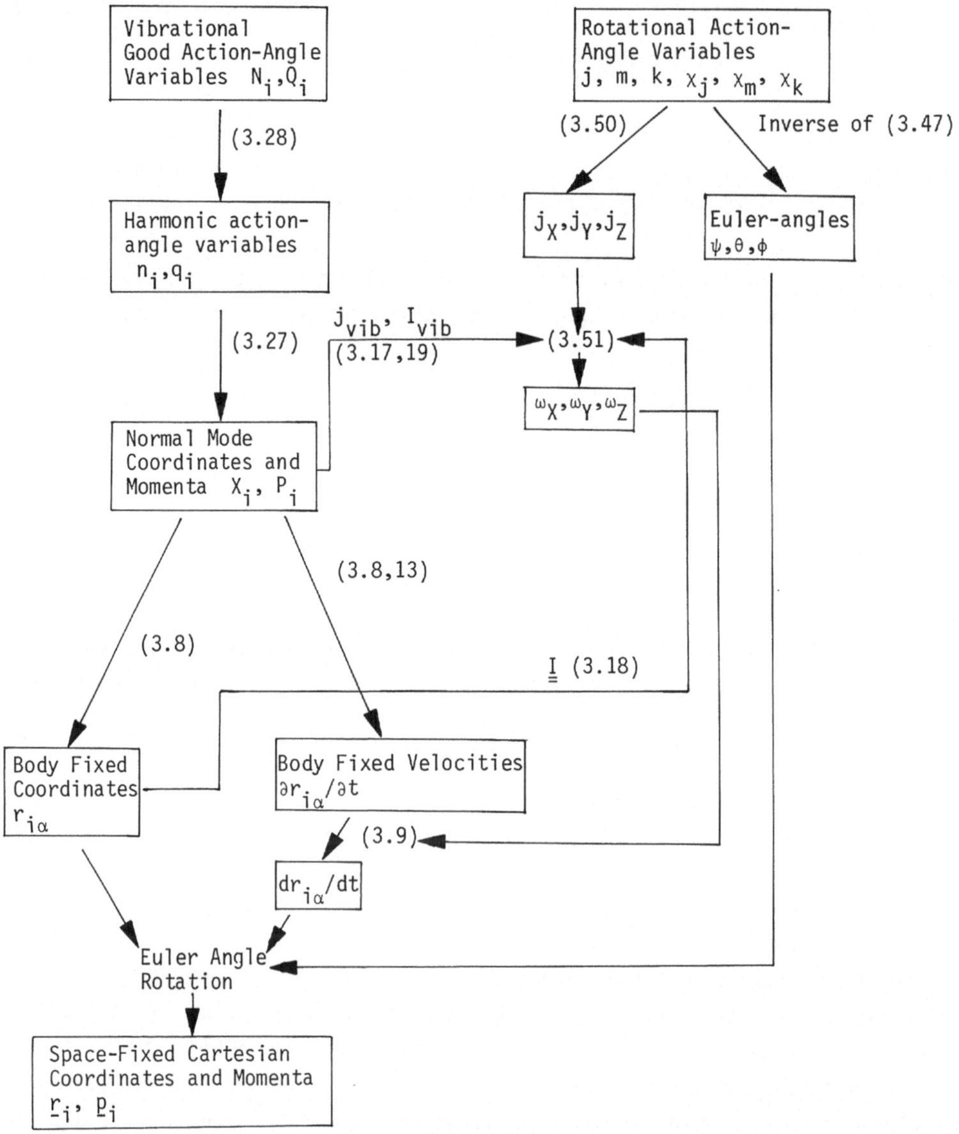

city transformations in (3.9,13), are obtained from the body-fixed angular momenta j_X, j_Y and j_Z using the following expression:

$$\omega = (\underline{\underline{I}} - \underline{\underline{I}}_{vib})^{-1}(\underline{j} - \underline{j}_{vib}) \quad , \tag{3.51}$$

where $\underline{j}$, $\underline{j}_{vib}$, $\underline{\underline{I}}$ and $\underline{\underline{I}}_{vib}$ are defined in (3.16-19), respectively. Note that in evaluating (3.51), $\underline{j}_{vib}$ and $\underline{\underline{I}}_{vib}$ are obtained from the normal coordinates and momenta. The evaluation of $\underline{\underline{I}}$ requires knowledge of the $r_{i\alpha}$'s as obtained from (3.8).

3.3.2 Trajectory Final Conditions

At the completion of each trajectory, the process outlined in Table 3.2 must be reversed to determine the action variables associated with each colliding molecule. If semiclassical S-matrix theory is used to determine transition probabilities, other information such as the good angle variables is also needed.

Table 3.3 summarizes the final action calculation. The rotational actions j and m can be calculated directly from the cartesian variables using (3.16), while k can be obtained using (3.16) after the Euler angle transformation from space-fixed to body-fixed coordinates. These Euler angles are calculated using (3.2) as discussed previously. Equation (3.4) is used to convert the body-fixed coordinates to normal coordinates X_k while (3.12) is used to define the angular velocities ω_X, ω_Y and ω_Z. Equation (3.9) then is used to convert from space to body time derivatives. Equation (3.13) and the time derivative of (3.4) then determine normal momenta P_k. Equations (3.27) then convert X_k, P_k to harmonic action-angle variables, and (3.28a) can be solved iteratively to determine the vibrational good actions N_k.

3.3.3 Assignment of Final Quantum States

Since the collisions are assumed to be governed by classical mechanics, there is nothing which constrains the final actions to be integral. This means that in order to compute state to state cross sections and rate constants, some sort of algorithm must be provided for relating the nonintegral final actions to quantum states. Although such algorithms have been extensively studied for diatomic systems, their use has rarely been considered for polyatomic systems. Some of the approaches developed for diatomics can, however, be generalized to polyatomics with little modification.

Semiclassical S-matrix theory, for example, uses those trajectories which satisfy semiclassical quantization conditions both initially and finally to calculate elements of the scattering matrix for specific transitions. Thus, as long as molecular vibrational motion is not chaotic in these two limits, such trajectories can be identified using the methods of the previous two sections. Unfortunately there are problems with this approach which make its usefulness limited in applications to polyatomics. These include (a) the computational difficulty associated with the multidimensional search for trajectories satisfying the proper boundary conditions, (b) the lack of simple but accurate and general algorithms for combining together the contributions to the scattering amplitude from multiple trajectories satisfying the boundary conditions [3.41], and (c) the difficulties associated with integrating complex trajectories to determine cross sections for forbidden transitions.

Table 3.3. Schematic diagram of final action determination process for polyatomic molecules

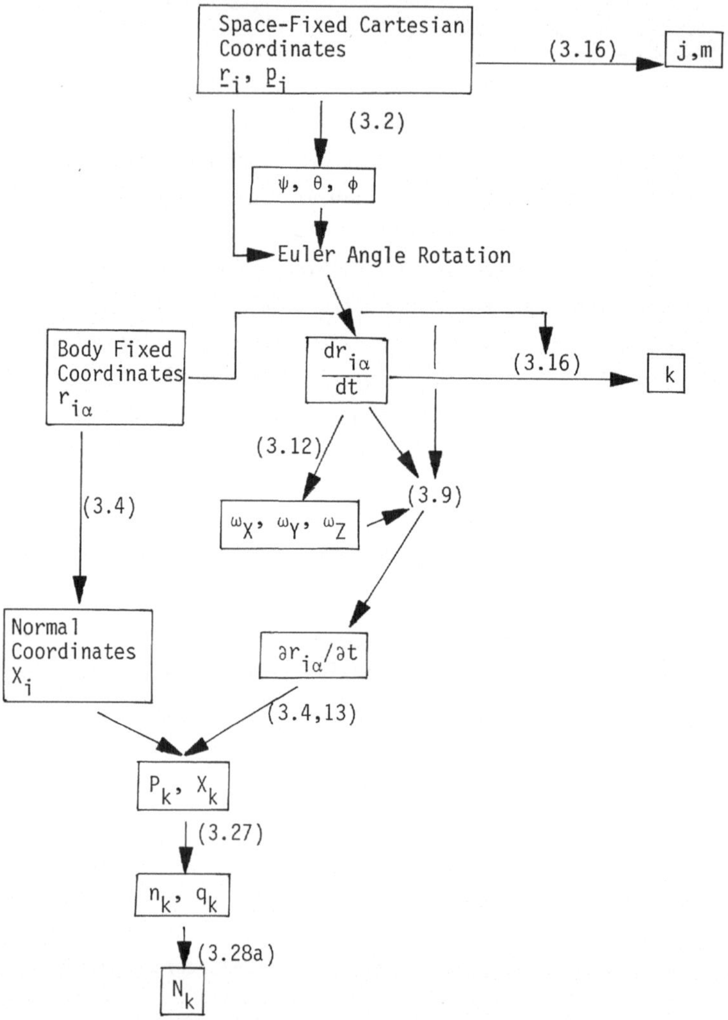

A much simpler approach to the assignment of quantum states to trajectories is the *classical histogram method*; this one simply rounds off the actions to the nearest integer to assign a quantum state to each trajectory. This approach has been thoroughly studied for diatomic systems and has been found to be reasonably accurate in determining state to state cross sections when many states are classically allowed. No analogous studies of this method for polyatomic systems have been done although there are only a few reasons to believe that the conclusions would be different. Perhaps one reason would arise when the final molecular motion is quasiperiodic and the rounded off actions correspond to regions of phase space for which trajectory motion is chaotic. Obviously the state assignment in this case is ambiguous. Another uncertain issue concerns the role of mode correlations in influencing vibrational distributions. Simply rounding off each action to the nearest integer independently of other actions ignores all mode correlation effects. It may be, however, that energy gap or other considerations favor a scheme in which the rounded actions of one mode might be functions of the actions of other modes. MILLER and MEYER [3.42], for example, used an algorithm of this sort in their studies of $F(^2P_{1/2}) + H_2(j) \rightarrow F(^2P_{3/2}) + H_2$. It should also be mentioned that threshold errors might be more severe for polyatomic than diatomic systems. This refers to transitions which are energetically forbidden but for which the rounding process (which typically ignores total energy constraints) makes it classically allowed. Since this can cause a cross section to become nonzero at energies as much as half a vibrational quantum *per mode* below threshold, it should be clear that for many mode systems, the associated error can be quite substantial. It is possible to partially correct for this by imposing total energy conservation in the histogram algorithm, although this will not help for transitions which are dynamically forbidden at energies well above their energetic threshold.

Another class of quantization schemes is based on the observation that the low-order moments of classical and quantum final state distributions for many diatomic problems are identical. These so called "*moment methods*" (including a number of forced oscillator methods such as INDECENT) have been successfully applied to several problems in diatomic molecule collisional energy transfer [3.43] (including the correct prediction of probabilities for classically forbidden processes), but only one study of its usefulness for polyatomic systems has been done. This involved a collinear model of Kr + CO_2, where CLARY [3.44] compared quantum moments for changes in symmetric and asymmetric stretch quantum numbers with the corresponding classical moments of SCHATZ and MULLONEY [3.16]. The extent of agreement was mixed, with the symmetric stretch first moments in generally good agreement but the asymmetric stretch moments of opposite sign. No comparisons of second moments or cross moments were made by him, but such information is available for a related model of Kr + CO_2 (one in which a harmonic potential was used for CO_2) by combining the quantum results of BOWMAN and LEASURE [3.5a] with those of SCHATZ and MULLONEY [3.17]. The results are given in Table 3.4 where we list the moments $\langle\Delta N_1\rangle$, $\langle\Delta N_1^2\rangle$, $\langle\Delta N_3\rangle$, $\langle\Delta N_3^2\rangle$ and $\langle\Delta N_1\Delta N_3\rangle$ for both Kr + CO_2 (000) and Kr + CO_2 (001). The comparison of first moments ($\langle\Delta N_1\rangle$ and $\langle\Delta N_3\rangle$) is analogous to that presented by CLARY, although it should be noted that for Kr + CO_2 (001), the $\langle\Delta N_3\rangle$ values are in reasonable agreement. The comparison of second and cross moments is fair for $\langle\Delta N_1^2\rangle$ but poor for $\langle\Delta N_3^2\rangle$ and $\langle\Delta N_1\Delta N_3\rangle$ for Kr + CO_2 (000). For (001), the second and cross moments are all in reasonable agreement. Note also that for Kr + CO_2 (001), $\langle\Delta N_1\Delta N_3\rangle$ is comparable to $\langle\Delta N_1^2\rangle$ and $\langle\Delta N_3^2\rangle$, while for (000) it is much smaller. This indicates that mode correlations are important for Kr + CO_2 (001)

Table 3.4. Comparison of quasiclassical and quantum moments for collinear Kr + CO_2[a]

A. Initial State = (000)

$E(eV)$[b]	$E_0(eV)$[c]	$<\Delta N_1>$	$<\Delta N_3>$	$<\Delta N_1^2>$	$<\Delta N_3^2>$	$<\Delta N_1 \Delta N_3>$
0.904	0.667	0.10(-4) 0.52(-5)	-0.50(-5) 0.50(-8)	0.22(-4) 0.52(-5)	0.29(-5) 0.50(-8)	-0.43(-5) 0.51(-10)
1.07	0.833	0.33(-4) 0.34(-4)	-0.84(-5) 0.74(-9)	0.89(-4) 0.34(-4)	0.79(-5) 0.74(-9)	-0.10(-4) 0.25(-10)
1.232	1.000	0.14(-3) 0.14(-3)	-0.15(-4) 0.18(-7)	0.32(-3) 0.14(-3)	0.19(-4) 0.18(-7)	-0.22(-4) 0.31(-10)
1.400	1.163	0.47(-3) 0.49(-3)	-0.23(-4) 0.57(-8)	0.89(-3) 0.49(-3)	0.38(-4) 0.57(-8)	-0.43(-4) 0.11(-10)

B. Initial State = (001)

$E(eV)$[b]	$E_0(eV)$[c]	$<\Delta N_1>$	$<\Delta N_3>$	$<\Delta N_1^2>$	$<\Delta N_3^2>$	$<\Delta N_1 \Delta N_3>$
0.904	0.370	0.71(-5) 0.48(-5)	-0.51(-5) -0.25(-5)	0.43(-5) 0.94(-5)	0.11(-5) 0.25(-5)	-0.22(-5) -0.48(-5)
1.07	0.536	0.18(-4) 0.14(-4)	-0.12(-4) -0.76(-5)	0.14(-4) 0.26(-4)	0.35(-5) 0.76(-5)	-0.61(-5) -0.14(-4)
1.232	0.703	0.44(-4) 0.38(-4)	-0.26(-4) -0.18(-4)	0.46(-4) 0.62(-4)	0.11(-4) 0.18(-4)	-0.16(-4) -0.30(-4)
1.400	0.866	0.10(-3) 0.10(-3)	-0.55(-4) -0.38(-4)	0.15(-3) 0.14(-3)	0.29(-4) 0.38(-4)	-0.36(-4) -0.56(-4)

[a] The classical moments (upper entries) are phase averaged values of the change in symmetric stretch action (ΔN_1) and asymmetric stretch action (ΔN_3), in units where $\hbar = 1$. The quantum moments (lower entries) are derived from the results of LEASURE and BOWMAN [3.5a] and represent the average change in symmetric and asymmetric stretch quantum number.
[b] Total energy, including zero point energy of the Kr + CO_2 system.
[c] E_0 is the initial relative translational energy.

but not (000). Indeed, for (001), BOWMAN and LEASURE [3.5a] found that the dominant low energy transition was (001) → (200) (a multiquantum transition which couples both modes) while the dominant transition from (000) was (000) → (100) (a single quantum transition involving just one mode).

Based on these comparisons, it is not clear that a simple moment method can be developed for Kr + CO_2. This system may very well be extreme in that there is a relative low density of states in the energy region considered so this conclusion does not necessarily mean that moment methods are not at all possible for polyatomic systems. It is also possible that other types of moments might display a better correspondence between classical and quantum mechanics. One might, for example, diagonalize the second moment matrix, then apply a moment method in this diagonalized representation. Or one might consider moments of the total vibrational energy transfer. Our attempts to use such methods in studies of Kr + CO_2 have so far produced mixed results (sometimes good, sometimes not). Thus, it remains unclear whether simple moment methods can generally be applied to polyatomic systems.

3.4 Collisional Excitation in He + SO_2

In this section, the trajectory methods developed in the previous section are used in a study of vibrational and rotational excitation in the He + SO_2 system. We choose the relative translational energies in this study sufficiently high so that several state to state excitation cross sections are classically allowed. This allows us to use the histogram method to determine cross sections and avoids some of the problems described in the previous section.

Although there have been several experimental measurements of collisional relaxation rates in rare gas/SO_2 mixtures [3.45], no attempt to compare the present results with those will be made. This is because (a) the atom-molecule interaction potential is not known quantitatively and (b) the experimental conditions are such that many state to state cross sections are needed at collision energies where the transitions are classically forbidden. In addition, the state which is initially excited in laser-induced fluorescence experiments is not well characterized so there are some uncertainties in the interpretation of the experimental data.

Our goal in this section is to analyze two types of information about collisional energy transfer calculations. First, we want to analyze the distribution of states obtained in the collisional excitation process so as to characterize its important features. Second, we wish to compare cross sections obtained using the partitioned Hamiltonian of (3.23) with those from a rotational sudden treatment of (3.22). This rotational sudden treatment has been used elsewhere in polyatomic collisional excitation [3.5c,19] but for vibrational excitation processes has never been compared quantitatively with the results of more accurate calculations. Of course, the partitioning in (3.23) is still far from an exact treatment of the polyatomic vibration-rotation motions, but it does include the full Coriolis interaction. The sudden treatment, on the other hand, neglects all but zero j Coriolis effects.

3.4.1 Details of Trajectory Calculations

Aside from the special methods used to define initial and final conditions, the trajectory simulation was a standard Monte Carlo application. The usual criteria of energy and angular momentum conservation were checked for each trajectory, and with a step size of 2.4×10^{-16} s used in a 5th-order Adams-Moulton predictor-corrector algorithm, energy conservation to 5 significant figures was obtained.

The intramolecular potential for SO_2 was assumed to be independent of the distance to the colliding He atom and was taken as the quartic force field of KUCHITSU and MORINO [3.35]. The intermolecular potential was obtained following an empirical prescription due to SUZUKAWA et al. [3.7]. In this the potential is written as a sum of three pairwise interactions (He- S, He-O and He-O), with the He-S interaction taken as that between He and Ar[46], and the He-O as that between He and Ne[46]. While the potential as written is certainly a crude approximation to the correct potential, this form has been used quite successfully in a previous study [3.19] in determining qualitative information such as final state distributions.

The rotational sudden calculations were done following the method outlined in [3.19]. Although it is possible to extract rotational distributions and related information from this type of calculation as outlined in [3.11], this part of the calculation does not have to be done simultaneously with the excitation part, and we have omitted it here. This has the advantage of reducing the number of equations of motion from 18 (for the full treatment) to 8 (3 normal coordinates and momenta plus the relative translational coordinate and momentum). This reduces the computational effort by a factor of about 2 in comparison with integrating the full set of equations of motion.

In both the sudden and fully-coupled treatments, the classical perturbation treatment was carried to second order in the determination of good action variables. For zero initial rotational angular momentum in the triatomic, this leads to good actions which were constant to within 0.5% or better in the absence of collisional interaction. This is comparable to the results obtained in studies of Kr + CO_2 [3.17] where it was shown that the action moments defined this way were reliable as long as they were larger than 10^{-2} $\hbar$. As will be seen below, this should be adequate for the present purposes.

3.4.2 Trajectory Results

Considering He + SO_2 (000) with zero initial rotational angular momentum, the integral cross sections $Q(N_1\ N_2\ N_3)$ for excitation of SO_2 to final states $(N_1\ N_2\ N_3)$ are presented in Tables 3.5,6 at relative translational energies E_0 of 1.0 and 2.0 eV, respectively. The cross sections were obtained by averaging over initial orbital angular momenta from 0 to 150 $\hbar$ at 1.0 eV and 0 to 200 $\hbar$ at 2.0 eV. This corresponds to maximum impact parameters of 6.7 a_0 and 6.3 a_0, respectively. These impact parameters are large enough to converge the inelastic integral cross sections, but the elastic cross section is probably not reliable. Also listed in Tables 3.5,6 are the moments $\langle\Delta N_1\rangle$, $\langle\Delta N_2\rangle$, $\langle\Delta N_3\rangle$ for the three modes of SO_2, the variances σ_1, σ_2, σ_3 [where $\sigma_i = (\langle N_i^2\rangle - \langle N_i\rangle^2)^{\frac{1}{2}}$] the cross variances σ_{12}, σ_{13} and σ_{23} [where $\sigma_{ij} = (|\langle N_iN_j\rangle - \langle N_i\rangle\langle N_j\rangle|)^{\frac{1}{2}}$], the average change in translational energy $\langle\Delta E_0\rangle$, in vibrational energy $\langle\Delta E_v\rangle$, in rotational energy $\langle\Delta E_r\rangle$ and the average final rotational angular momentum $\langle j\rangle$. The results in each table are listed first from the coupled calculations [wherein the Hamiltonian in (3.23) was used in the vibrational action calculation], and then from the sudden rotation calculations [wherein trajectories

Table 3.5. Integral cross sections ($bohr^2$) and moments for He + SO_2 (000) → He + SO_2 ($N_1N_2N_3$) at E_0 = 1 eV and l_{max} = 150 $\hbar$[a]

Final state ($N_1N_2N_3$)	2nd order energy [cm^{-1}]	Integral cross section Coupled	Sudden
000	1528.9	125.1 ± 2.7[b]	123.9 ± 2.6
010	2044.7	13.0 ± 1.4	10.3 ± 1.2
020	2554.5	1.8 ± 0.5	2.6 ± 0.6
100	2684.1	0.6 ± 0.2	0.8 ± 0.4
110	3197.9	0.2 ± 0.2	0.2 ± 0.1
120	3705.6	0.1 ± 0.1	0.4 ± 0.4
$\langle \Delta N_1 \rangle$	0.002	0.010	
$\langle \Delta N_2 \rangle$	0.105	0.122	
$\langle \Delta N_3 \rangle$	0.007	0.002	
σ_1	0.099	0.085	
σ_2	0.354	0.415	
σ_3	0.068	0.056	
σ_{12}	0.017	0.058	
σ_{13}	0.035	0.040	
σ_{23}	0.041	0.026	
$\langle \Delta E_0(eV) \rangle$	-0.038	-0.010	
$\langle \Delta E_v(eV) \rangle$	0.008	0.010	
$\langle \Delta E_r(eV) \rangle$	0.030	-	
$\langle j \rangle$	15.8	-	

[a] All action moments are in units of $\hbar$. These results were obtained using 500 trajectories. Statistical uncertainties given represent plus or minus one standard deviation.
[b] Elastic cross section is not accurate (see text).

at fixed atom-molecule orientations are integrated and no rotational energy transfer information is obtained].

Let us first analyze the final state distributions in Tables 3.5,6. These distributions are plotted in Figs.3.1,2 as a function of the vibrational energy E_v (N_1 N_2 N_3) associated with each final state. These energies [obtained from the 2nd order expression (3.42)] are listed in Tables 3.5,6. Note that the states listed in these two tables are all those for which nonzero histogram cross sections were calculated using the coupled method with 500 and 1000 trajectories, respectively. Not surprisingly, there are many more states listed at 2 eV than at 1 eV. States for which the histogram cross section is zero, but which have E_v in the range considered in Figs.3.1,2, are indicated by an X on the abscissa.

There are many ways to analyze the final state distributions in Tables 3.5,6 and Figs.3.1,2. For example, an often used characterization [3.47] concerns the gap ΔE in energy from the initial vibrational state to the final one. In many studies of diatomic molecule collisional energy transfer [3.47] one finds that the cross section decreases exponentially with increasing $|\Delta E|$. In the present case, ΔE is ap proximately equal to $E_v(N_1\ N_2\ N_3)$ - $E_v(000)$ and thus the lines in Figs.3.1,2 should decrease in length exponentially with $E_v(N_1\ N_2\ N_3)$ if the exponential law is to be valid. In actual fact, Figs.3.1,2 do show a general decrease in length with in-

Table 3.6. Integral cross sections ($bohr^2$) and moments for He + $SO_2(000) \rightarrow$ He + $SO_2(N_1N_2N_3)$ at E_o = 2 eV and l_{max} = 200 ℏ[a]

Final state ($N_1N_2N_3$)	2nd order energy [cm^{-1}]	Integral cross section Coupled	Sudden
000	1528.9	103.7 ± 2.0[b]	103.1 ± 2.2[b]
010	2044.7	10.6 ± 1.0	11.5 ± 1.0
020	2554.5	3.8 ± 0.6	2.4 ± 0.4
030	3058.3	0.9 ± 0.2	1.8 ± 0.3
040	3556.2	0.5 ± 0.2	0.8 ± 0.2
050	4048.1	0.3 ± 0.1	0.4 ± 0.1
060	4534.0	0.1 ± 0.1	0.2 ± 0.1
070	5013.9	0.0 ± 0.1	0.1 ± 0.1
100	2684.1	2.7 ± 0.4	1.2 ± 0.2
110	3197.9	0.7 ± 0.2	0.6 ± 0.2
120	3705.6	0.5 ± 0.2	0.3 ± 0.2
130	4207.3	0.2 ± 0.1	0.3 ± 0.2
140	4703.1	0.1 ± 0.1	0.1 ± 0.1
150	5192.9	0.1 ± 0.1	0.0 ± 0.1
200	3831.5	0.3 ± 0.1	0.1 ± 0.1
210	4343.1	0.1 ± 0.1	0.0 ± 0.1
220	4848.8	0.1 ± 0.1	0.0 ± 0.1
300	4970.9	0.0 ± 0.1	0.0 ± 0.1
001	2887.6	0.4 ± 0.1	1.2 ± 0.8
011	3399.5	0.1 ± 0.1	0.3 ± 0.1
021	3905.4	0.0 ± 0.1	0.2 ± 0.1
101	4029.1	0.2 ± 0.1	0.1 ± 0.1
$\langle\Delta N_1\rangle$	0.043	0.023	
$\langle\Delta N_2\rangle$	0.219	0.254	
$\langle\Delta N_3\rangle$	0.084	0.012	
σ_1	0.246	0.209	
σ_2	0.676	0.806	
σ_3	0.125	0.136	
σ_{12}	0.125	0.151	
σ_{13}	0.010	0.060	
σ_{23}	0.038	0.087	
$\langle\Delta E_o(eV)\rangle$	-0.073	-0.022	
$\langle\Delta E_v(eV)\rangle$	0.021	0.022	
$\langle\Delta E_r(eV)\rangle$	0.051	-	
$\langle j\rangle$	20.7	-	

[a]Notation is analogous to that in Table 3.5. These results are for 1000 trajectories.
[b]Elastic cross section is not accurate.

creasing E_v, but this decrease is not exponential, and more importantly, the distribution shows mode specificity in that the decay for some modes is faster than for others. For example, at E_o = 2 eV the excited asymmetric stretch state (001) with E_v = 2887.6 cm^{-1} has a cross section (0.4 a_o^2) which is much lower than for states which have similar energies (such as 020, and 100) but which involve different modes. Thus, one cannot characterize these distributions solely based on energy gap consideration.

Another characterization of these final state distributions is based on the quantum number changes associated with each vibrational mode. For example, it is clear from Table 3.6 that the cross sections decrease rapidly with increasing N_2

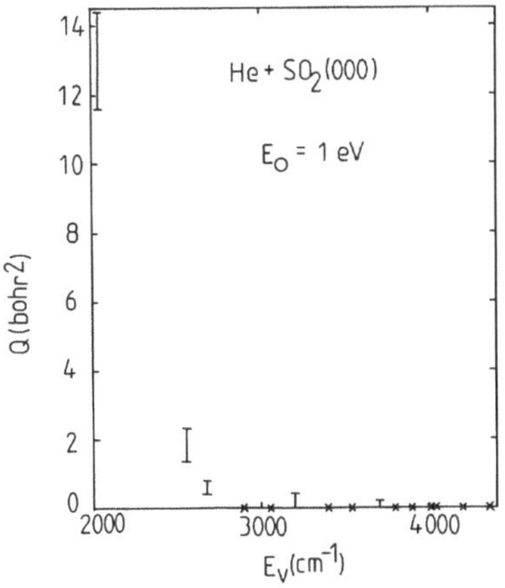

Fig.3.1

Fig.3.2

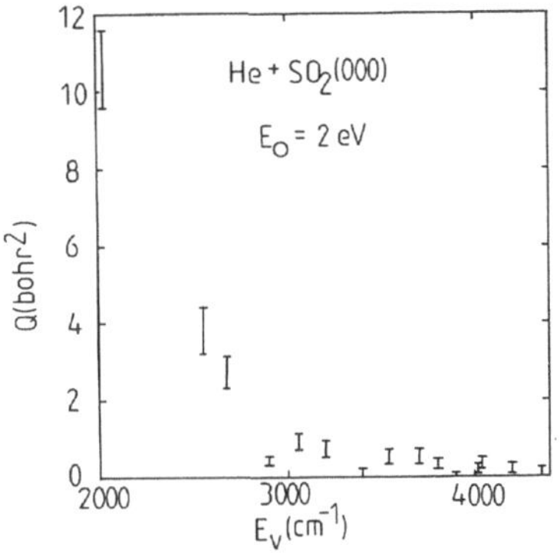

Fig.3.1. Integral cross section Q [$bohr^2$] versus final state vibrational energy E_v for He + SO_2(000) → He + SO_2(N_1 N_2 N_3) at E_0 = 1 eV (results from Table 3.5). The error bars indicate ± one standard deviation about the average value of the cross section. An X in the abscissa identifies a state for which the histogram cross section is zero. The final states having nonzero cross sections are (from left to right): (010), (020), (100), (110) and (120). Those having zero cross section are: (001), (030), (011), (040), (200), (021), (101), (050), (130) and (210)

Fig.3.2. Integral cross section Q versus E_v as in Fig.3.1 but for E_0 = 2 eV. The final states indicated are (from left to right): (010), (020), (100), (001), (030), (110), (011), (040), (120), (200), (021), (101), (050), (130) and (210)

for a given N_1, N_3. [Consider, for example the sequence (000), (010), (020), (030),...]. The rate of decrease is of course different for different modes (this is best seen by comparing the variances σ_1, σ_2 and σ_3). If the modes were uncorrelated, the rate of decrease of one mode would not depend on the actions associated with the other two. Unfortunately, this latter statement does not work for the distributions in Tables 3.5,6, for the correlations evident in σ_{12}, σ_{13}, and σ_{23} are only somewhat smaller than the variances σ_1, σ_2 and σ_3.

The overall characterization of the distributions pictured in Figs.3.1,2 is then somewhat between the uncorrelated mode limit and the energy gap limit. Moreover, the importance of mode correlations seems to change substantially with E_0. At E_0 = 1.0 eV, σ_{13} and σ_{23} are comparable and larger than σ_{12} (in the coupled results), while at 2 eV, $\sigma_{12} > \sigma_{23} > \sigma_{13}$. Other calculations we have done indicate that modes 1 and 3 show the strongest correlation at E_0 much less than 1.0 eV, and that modes 1 and 2 are strongly correlated at high E_0. The low energy behavior seems to be governed by frequency differences, with the strongest correlation between modes with the smallest frequency difference (in SO_2, ω_1 = 1171.1 cm^{-1}, ω_2 = 524.8 cm^{-1}, and ω_3 = 1377.8 cm^{-1}). At high E_0, the strongest correlations are between the most easily excited modes. In this case, these are the lowest frequency modes.

The energy moments in Tables 3.5,6 also provide much useful information about the dynamics. Considering the "coupled" results at both 1 and 2 eV, the average translational energy transfer is negative and both the rotational and vibrational energy transfers are positive. In addition, $<\Delta E_r>$ is significantly larger than $<\Delta E_V>$, indicating that T → R energy transfer predominates over T → V. The ratio T → V/T → R does increase with increasing E_0, however. Note also that the $<\Delta E_V>$'s from the sudden calculations are quite similar to those from the coupled results. This suggests that the separation of vibrational energy transfer from rotation is an excellent approximation, and further evidence for this is documented below. Because the sudden treatment neglects rotational energy transfer, the sudden $<\Delta E_0>$ is much smaller than the coupled value (by energy conservation, $<\Delta E_0> + <\Delta E_V> + <\Delta E_r> = 0$). A sudden value for $<\Delta E_r>$ could, however, be calculated using the methods outlined in [3.11].

Comparing now the coupled and sudden cross sections in Tables 3.5,6, we find that the results for most transitions are in good agreement. All but one cross section in Table 3.5 and all but three in Table 3.6 agree to within the sum of the standard deviations given. The action moments are likewise very similar, although the sudden values of $<\Delta N_2>$ and σ_2 in both tables are systematically larger than the coupled results. This suggests that there is slightly more bend mode excitation in the sudden results. In any event, the excellent agreement of the sudden and coupled results indicates the validity of the sudden treatment for He + SO_2, at 1-2 eV translational energy. It also means that the Coriolis effects which are included in the coupled treatment of semiclassical eigenstates but not in the sudden one are not particularly important to the cross sections presented.

3.5 Conclusion

In this paper a detailed analysis of the use of the quasiclassical trajectory method in studies of polyatomic molecule nonreactive collisions has been presented. Unlike the analogous quantum mechanical treatments (where the initial and final state specification is straightforward, but the collision dynamics calculation is very difficult), the trajectory method provides a relatively simple approach to describing collision dynamics but is complicated by difficulties in describing initial and final semiclassical molecular eigenstates. The zeroth-order treatment (rigid rotor-harmonic oscillator) is simple enough, but is often inapplicable because Coriolis and anharmonic effects cause the molecular Hamiltonian to be nonseparable. The approaches outlined in this paper for partitioning this Hamiltonian between vibration and rotation and for solving the vibrational and rotational Hamilton-Jacobi equation are based on simple perturbation solutions using the zeroth-order states as reference. This is adequate for molecules with small amounts of vibrational and rotational excitation and as such is useful for describing many problems dealing with thermal relaxation phenomena. Even then, however, technology at present only allows for an accurate treatment of triatomic molecules, and particularly those states of triatomics which are not strongly perturbed by Fermi resonance. Of course, generalizations and improvements are possible. States perturbed by Fermi resonance can be treated by degenerate perturbation theory methods [3.24] and molecules larger than triatomics are amenable to study using the same methods as described here, admittedly with significantly more computational effort. The treatment of highly rotationally and/or vibrationally excited molecules is a serious problem, however, even for triatomics, as the fundamental interrelationship between quantum and classical mechanics for chaotic molecular motion is unclear. Since there are a growing number of experimentally relevant chemical systems where collisions with highly excited molecules are involved, the elucidation of this classical/quantum correspondence is an important topic for future research. The results of our application to He + SO_2 illustrate some of the wealth of information which can be obtained from quasiclassical calculations once the above described problems with initial and final state specification are overcome. Particularly, we found that the final SO_2 vibrational distributions are not amenable to simple characterizations in terms of uncorrelated normal modes or exponential gap distributions. Rather, these distributions demonstrate that the interplay between the many dynamical factors (frequency differences, translational, vibrational and rotational timescales, anharmonic couplings, etc.) present in polyatomic systems can give complex final state distributions. Obviously, much is yet to be learned about polyatomic molecule collision processes before our understanding of such systems begins to approach that currently available for diatomics.

Acknowledgement. I gratefully acknowledge support from NSF grant CHE-7820336. This research was supported in part by the National Resource for Computation in Chemistry under a grant from the National Science Foundation and the U.S. Department of Energy (contract No. W-7405-ENG-48).

It is a pleasure to acknowledge numerous discussions concerning this research with Prof. Joel Bowman.

References

3.1 W. Eastes, U. Ross, J.P. Toennies: J. Chem. Phys. *66*, 1919 (1977); J. Krutein, F. Linder: J. Phys. B*10*, 1363 (1977)

3.2a E. Weitz, G. Flynn: Ann. Rev. Phys. Chem. *25*, 275 (1974);

3.2b G.T. Fujimoto, E. Weitz: Chem. Phys. *27*, 65 (1978)

3.3 R.N. Schwartz, Z.I. Slawsky, K.F. Herzfeld: J. Chem. Phys. *20*, 1591 (1952); R.N. Schwartz, K.F. Herzfeld: J. Chem. Phys. *22*, 767 (1954); F.I. Tanczos: J. Chem. Phys. *25*, 439 (1956); J.L. Stretton: Trans. Far. Soc. *61*, 1053 (1965); A. Miklavc, S. Fischer: Chem. Phys. Lett. *44*, 209 (1976)

3.4a M.E. Coltrin, R.A. Marcus: J. Chem. Phys. *73*, 2179 (1980);

3.4b J.W. Duff, N.C. Blais, D.G. Truhlar: J. Chem. Phys. *71*, 4304 (1979);

3.4c R.N. Porter: Ann. Rev. Phys. Chem. *25*, 317 (1974), and references therein

3.5a J.M. Bowman, S.C. Leasure: Chem. Phys. Lett. *56*, 183 (1978);
3.5b D.A. Micha: J. Chem. Phys. *70*, 565 (1979);
3.5c D.C. Clary: Mol. Phys. *39*, 1295 (1980);
3.5d D.C. Clary: J. Chem. Phys. *75* , 209 (1981);
3.5e D.C. Clary: J. Chem. Phys. *75*, 2899 (1981)
3.6 R.L. Thommarson, G.C. Berend, S.W. Benson: J. Chem. Phys. *54*, 1313 (1971)
3.7a H.H. Suzukawa: Ph. D. Thesis, University of California at Irvine, 1974;
3.7b H.H. Suzukawa, M. Wolfsberg, D.L. Thompson: J. Chem. Phys. *68*, 455 (1978)
3.8 G.C. Schatz, M.D. Moser: J. Chem. Phys. *68*, 1992 (1978)
3.9 N. Sathymurthy, L.M. Raff: J. Chem. Phys. *66*, 2191 (1977)
3.10 A.J. Stace, J.N. Murrell: J. Chem. Phys. *68*, 3028 (1978)
3.11 T. Mulloney, G.C. Schatz: Chem. Phys. *45*, 213 (1980)
3.12 J.D. MacDonald, R.A. Marcus: J. Chem. Phys. *65*, 2180 (1976)
3.13 W.L. Hase, R.J. Wolf, C.S. Sloane: J. Chem. Phys. *71*, 2911 (1979), and references therein
3.14 S. Chapman, D.L. Bunker: J. Chem. Phys. *62*, 2890 (1975)
3.15 G.D. Billing: Chem. Phys. *33*, 227 (1978); Chem. Phys. *46*, 123 (1980)
3.16 G.C. Schatz, T. Mulloney: J. Chem. Phys. *71*, 5257 (1979)
3.17 G.C. Schatz, T. Mulloney: J. Phys. Chem. *83*, 989 (1979)
3.18 S.C. Chapman, B.C. Garrett, W.H. Miller: J. Chem. Phys. *64*, 502 (1976)
3.19 G.C. Schatz: J. Chem. Phys. *72*, 3929 (1980)
3.20 M.J. Redmon, R.J. Bartlett, B.C. Garrett, G.D. Purvis, P.M. Saatzer, G.C. Schatz, I. Shavitt: *Potential Energy Surfaces and Dynamics Calculations*, ed. by D.G. Truhlar (Plenum, New York 1981) p.771
3.21 G.C. Schatz, M.J. Redmon: Chem. Phys. *58*, 195 (1981)
3.22 G.C. Schatz: J. Chem. Phys. *71*, 542 (1979)
3.23 G.C. Schatz, H. Elgersma: Chem. Phys. Lett. *73*, 21 (1980);
G.C. Schatz: J. Chem. Phys. *74*, 1133 (1981)
3.24 M. Born: *The Mechanics of the Atom* (G. Bell and Sons, London 1927)
3.25a J. Ford: Adv. Chem. Phys. *24*, 155 (1973);
3.25b V.I. Arnold, A. Avez: *Ergodic Problems of Classical Mechanics* (Benjamin, New York 1968);
3.25c I.C. Percival: Adv. Chem. Phys. *36*, 1 (1977);
3.25d D.W. Noid, M.L. Koszykowski, R.A. Marcus: J. Chem. Phys. *71*, 2864 (1979), and references therein
3.26 G. Contopoulos: Bull. Astron. *2*, 223 (1967)
3.27 C. Jaffe, W.P. Reinhardt: J. Chem. Phys. *71*, 1862 (1979)
3.28 G.C. Schatz: Chem. Phys. Lett. *67*, 248 (1979)
3.29 G.C. Schatz, M.D. Moser: Chem. Phys. *35*, 239 (1979)
3.30a E.B. Wilson, J.C. Decius, P.C. Cross: *Molecular Vibrations* (McGraw-Hill, New York 1955)
3.30b S. Califano: *Vibrational States* (Wiley, New York 1976)
3.31 H. Elgersma, G.C. Schatz: Chem. Phys. *54*, 201 (1980)
3.32 R.N. Porter, L.M. Raff, W.H. Miller: J. Chem. Phys. *63*, 2214 (1975)
3.33 Program KC05, *NRCC Software Catalog* (Vol. 1), Lawrence Berkeley Laboratory (Document LBL-10811) (1980)
3.34 N.C. Handy, S.M. Colwell, W.H. Miller: Far. Disc. Chem. Soc. *62*, 29 (1977)
3.35 K. Kuchitsu, Y. Morino: Bull. Chem. Soc. Japan *38*, 814 (1965)
3.36 A.R. Hoy, I.M. Mills, G. Strey: Mol. Phys. *24*, 1265 (1972)
3.37 R.J. Whitehead, N.C. Handy: J. Mol. Spect. *55*, 356 (1975)
3.38 S. Augustin, W.H. Miller: J. Chem. Phys. *61*, 3158 (1974)
3.39 S.M. Colwell, N.C. Handy, W.H. Miller: J. Chem. Phys. *68*, 745 (1978)
3.40 D.L. Bunker: Meth. Comp. Phys. *10*, 287 (1971)
3.41 C.W. McCurdy, W.H. Miller: J. Chem. Phys. *73*, 3191 (1980)
3.42 H.D. Meyer, W.H. Miller: J. Chem. Phys. *71*, 2156 (1979)
3.43 D.G. Truhlar, J.W. Duff: Chem. Phys. Lett. *36*, 551 (1975)
3.44 D.C. Clary: J. Chem. Phys. *75*, 2023 (1981)
3.45 D. Siebert, G. Flynn: J. Chem. Phys. *62*, 1212 (1975);
R.C. Slater, G.W. Flynn: J. Chem. Phys. *65*, 425 (1976)
3.46 C.H. Chen, P.E. Siska, Y.T. Lee: J. Chem. Phys. *59*, 601 (1973)
3.47 J.C. Polanyi, K.B. Woodhall: J. Chem. Phys. *56*, 1563 (1972);;
R.B. Bernstein: J. Chem. Phys. *62*, 4570 (1975)

4. Rotational Rainbows in Atom-Diatom Scattering

R. Schinke and J.M. Bowman

With 20 Figures

Systematic structures in cross sections are always very appealing for the scientist working in the field of molecular collisions. We mean those structures which vary in a *predictable* way as the parameters determining the collision conditions (i.e., energy, masses, scattering angle, coupling strength,...) are changed. The rainbow structures in the differential cross section and the glory oscillations in the integral cross section are important examples for isotropic potential scattering. Usually a deep physical understanding of the collision dynamics is gained from the detailed analysis of such structures which, in turn, makes the determination of the interatomic potential from measured quantities, in principle, feasible. This has been amply demonstrated for atom-atom scattering.

In recent years a new type of structure in rotationally inelastic cross sections has been predicted by theory and later observed in a number of nice experiments, i.e., *rotational rainbows*. They are closely related to the rainbow effect in atom-atom scattering and it is tempting to speculate whether they will play a similar role in the determination of atom-molecule potential energy surfaces. It is the purpose of this chapter to review the recent developments in rotational rainbow scattering.

Related structures might have been observed in experimental or, more likely, in theoretical studies *before* they had been recognized as rainbows. We do not attempt to give a complete list of these earlier studies. Full account will be made only of the theoretical and experimental investigations which describe or explain the new structures as rainbows.

4.1 Background

4.1.1 Experiment

Rotational energy transfer is one of the fundamental processes in the gas phase. We consider the simple case where a structureless atom collides with a diatomic molecule. The potential energy surface $V(R,r,\gamma)$ depends on three variables usually taken as the distance between the molecular center-of-mass and the atom R, the intermolecular separation r, and the orientation angle γ, formed by both vectors.

Roughly speaking, the r and γ-dependence of V induce vibrationally and rotationally inelastic transitions, while the R-dependence determines the angular behavior of the measurable quantities, the differential cross sections. For neutral systems, vibrationally inelastic processes are normally very weak for collision energies below 1 eV or so, and the scattering is well described in the so-called rigid-rotor approximation with r fixed at the asymptotic equilibrium distance r_{eq} and $V = V(R,\gamma)$ (another possibility is to average V over r, of course). Thus, encounters between an atom and a rigid-rotor diatom are the next more complicated processes after scattering from an isotropic potential V(R).

In order to obtain a full understanding of the collision dynamics and detailed information about the potential energy surface, differential cross sections have to be measured as a function of the scattering angle θ *and* the rotational transition $j_1 \rightarrow j_2$, i.e., $d\sigma/d\Omega(j_1 \rightarrow j_2|\theta)$. Transitions between different magnetic states m_1 and m_2 will not be considered because all experimental and theoretical cross sections discussed in this chapter are summed and averaged over these states. Because of substantial experimental difficulties, fully state resolved differential cross sections have been reported for only a few systems in recent years [4.1,2].

The neutral systems for which measurements of rotationally inelastic differential cross sections have been made can be roughly divided into two groups. The first group (I) contains Ne-HD [4.3], Ne-D_2 [4.4], He-HD [4.5], D_2-HD [4.6], and HD-HD [4.7]. All of these atom-diatom or diatom-diatom measurements are characterized by relatively low initial translational energies E and light target molecules with large rotational constants B_{rot} such that the ratio E/B_{rot} is typically small ($\lesssim$ 10). Thus, only a few rotational transitions are energetically accessible. As a consequence, the elastic process is mainly dominant over a wide range of scattering angles and the inelastic transitions play only a minor role. The features known from atom-atom scattering typically govern both the elastic and the inelastic differential cross sections. Because of the small wave numbers, these are diffraction oscillations due to scattering from the repulsive part of the potential. Rainbow structures due to interference of "trajectories" with different impact parameters leading to the same scattering angle θ are hardly observed in these experiments because the well depths are much smaller than the collision energy. The recent state resolved measurements for He-N_2, CO and CH_4 [4.8] seem to belong to this group (I), although the ratio E/B_{rot} is of the order of 100.

The second group (II) contains the systems K-N_2 [4.9], K-CO [4.9,10], He-Na_2 [4.11,12], Ne-Na_2 [4.13,14] and Ar-Na_2 [4.13,15], Kr-Na_2 and Xe-Na_2 [4.13.b]. All of these experiments are characterized by relatively high collision energies and heavy target molecules with small rotational constants such that the ratio E/B_{rot} is typically large ($\gtrsim$ 1000) and many excited rotational states are energetically accessible. Because of the large value of the wave numbers, the diffraction oscillations are very narrow and restricted to small scattering angles; they have not been resolved in these experiments. In addition, rainbow scattering as known from atom-atom scattering is unimportant because the well depths of the respective potentials are again negligibly small compared to the collision energy (Sect.4.5). Instead, a new type of systematic structure is observed in the angular distribution of Δj differential cross sections as illustrated in Fig.4.1 showing experimental and theoretical $j_1 = 0 \to 2$, 4 and 6 differential cross sections for Ne-Na_2 [4.14]. Let us first discuss the $0 \to 6$ cross section. It rises steeply out of the forward direction, reaches a maximum at an angle of about $\theta_{LAB} \simeq 5^\circ$ ($\theta_{CM} \simeq 15^\circ$) and declines to larger scattering angles where it shows two additional maxima. This behavior is expected to be typical for any inelastic differential cross section for group II systems. The main maxima are off scale for $j_2 = 2$ and 4. They shift to larger scattering angles as j_2 is increased. The spacing of the angular oscillations is roughly $\Delta\theta_{LAB} \simeq 7^\circ$ ($\Delta\theta_{CM} \simeq 20^\circ$) for $j_2 = 6$, which has to be compared with $\Delta\theta_{CM} \simeq 0.7^\circ$ as one would expect for the diffraction oscillations at the energy of $E = 175$ meV [4.16].

Structures in j_2-distributions at fixed scattering angle, which are closely related to the ones shown in Fig. 4.1, have been reported in [4.9,10,12,17]. We anticipate that the structures in rotationally inelastic differential cross sections discussed in Fig.4.1 for Ne-Na_2 will be observable for *any* system provided the collision energy is high enough to ensure that many rotational states are energetically accessible and the coupling is sufficiently strong to excite them. It is the purpose of this chapter to discuss the origin of these features and their relation to the potential energy surface and other collision parameters.

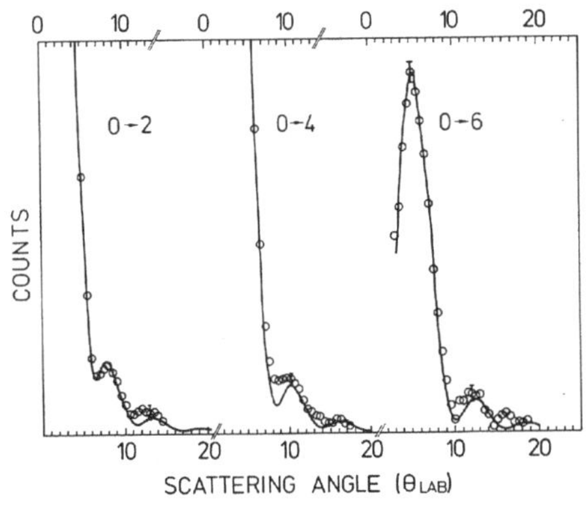

Fig.4.1. Experimental (ooo) and theoretical (——) $j_1 = 0 \to j_2 = 2$, 4 and 6 Ne-Na_2 differential cross sections versus laboratory scattering angle. The center-of-mass collision energy is 175 meV. Experimental and theoretical data are normalized to each other at *one* point, $\theta_{LAB} = 5^\circ$ for $j_2 = 6$ (adapted from HEFTER et al. [4.14])

4.1.2 Theory

There is a wide variety of theoretical methods that we might consider using to calculate rotationally inelastic differential cross sections. The method of choice ideally should be quantitative yet feasible to apply to the group II systems just discussed. Also, the method or simple approximations based on it should provide physical insight and understanding of the very interesting features in the differential cross sections shown in Fig.4.1.

The close-coupling method (also known as the coupled-channel method) is, in principle, an exact quantum method. It has recently been reviewed by several authors [4.18,19]and by Secrest in Chap.2. Due to the computational difficulties in implementing this method, it has been mainly applied to group I systems, i.e., those with relatively few coupled rotational states. Unfortunately, because of its prohibitive computational difficulty for group II systems we cannot consider it to be the method of choice for our present purpose. An important approximation to the close-coupling method is the so-called coupled-states method. This has also been recently reviewed [4.20] and again the Chap.2 by Secrest can be consulted for details of the method. This method can be (and has been) used to study some group II systems (He-Na_2 [4.21], Ne-Na_2 [4.16]; see also Sect.4.4.1). One drawback of the method is its considerable complexity, which renders it somewhat difficult to interpret physically.

The quasiclassical trajectory method can be used to study rotationally inelastic scattering. A review of this specific application of the method [4.22] and an assessment of its accuracy have been recently given. The conclusion that can be drawn is that the method is qualitatively correct, sometimes quantitatively accurate. The obvious limitation in not describing quantum effects such as tunneling and interference phenomena was noted. The method's assets are its relative ease of application and physical interpretation. Classical S-matrix theory is a powerful approximation which utilizes exact classical trajectories and in principle describes all quantum effects. Unfortunately, the method can be difficult to implement and its accuracy for rotationally inelastic scattering is not always quantitative. Certainly its great asset is its interpretative power which has been amply demonstrated [4.23,24].

The infinite order sudden (IOS) approximation (also called the sudden rotation approximation) is an important approximate quantum theory which is ideally suited for a study of group II systems, the ones of interest to us. The quantum theory has been recently reviewed, specifically for rotational excitation [4.20] (also, see the Chap.2). The basic assumption in the theory is that the characteristic time of rotational motion is much greater than the collision time. The inequality, $E/B_{rot} \gg 1$, is an equivalent way of expressing this time scale difference. Thus, roughly speaking, the accuracy of the sudden approximation is expected to increase as the number of energetically open rotational states increases. This of course is just the regime where the close-coupling and coupled-states methods are increasingly difficult to implement. Many computational studies of rotational excitation have been reported using the IOS method and its accuracy (and limitations) are well-documented (an extensive list of relevant references is given in [4.20]). Recently, the classical limit of the IOS theory of rotational excitation has been derived [4.25-27] and shown to be a very useful way to interpret the features of the differential cross sections shown in Fig.4.1 [4.21,25,27-29]. Thus, we will focus on the IOS theory and its classical limit in this chapter. We will, however, take note of important contributions made by others, using different theoretical methods to analyse the structures of rotationally inelastic differential cross sections.

4.1.3 Outline

In Sect.4.2 a review of the Ford-Wheeler analysis of atom-atom, isotropic potential scattering is given with emphasis on their classic rainbow analysis of the differential cross section. Following that in Sect.4.3 the quantum IOS approximation will be reviewed and its classical limit described in detail. The fundamental interpretation of maxima in rotationally inelastic differential cross sections as rotational rainbows will be made and their general properties illustrated for group II systems. A short discussion of the relationship between the classical limit IOS theory, classical S-matrix theory and classical theory concludes that section. Some numerical examples for atom-homonuclear and atom-heteronuclear diatom systems will be considered in Sect.4.4. Particular experiments which show rainbow structures will be discussed in Sect.4.5. There were also make a quantitative comparison between experiment and theory for the systems He, Ne-Na_2. A variety of simple analytical expressions for rough but easy interpretation of differential cross sections will be presented in Sect.4.6. These are based on either an approximate hard shell representation of the interaction potential but otherwise accurate IOS analysis, or an approximate IOS analysis but using realistic interaction potentials. A summary and prognosis for future work will be given in the concluding section.

4.2 Review of Rainbows in Isotropic Potential Scattering

This subject has recently been thoroughly reviewed by PAULY [4.30] so we will focus on those aspects of rainbow scattering by isotropic potentials which are especially germane to the theory which follows for rotationally inelastic and elastic scattering in atom-diatom collisions.

To begin, let us recall the familiar expression for the scattering amplitude $f(\theta)$ for central field potential scattering:

$$f(\theta) = \frac{i}{2k} \sum_{\ell} (2\ell + 1)[1 - \exp(2i\eta_{\ell})]P_{\ell}(\cos\theta) \quad , \tag{4.1}$$

where $P_{\ell}(\cos\theta)$ is the Legendre polynomial of order ℓ, η_{ℓ} is the phase shift and k, the wave number, is related to the asymptotic translational energy E by

$$k = (2\mu E)^{\frac{1}{2}}/\hbar \quad ,$$

where μ is the reduced mass of the system. The scattering observables are the differential and total cross sections given, respectively, by

$$\frac{d\sigma}{d\Omega}(\theta) = |f(\theta)|^2 \tag{4.2a}$$

$$\sigma = \int d\Omega |f(\theta)|^2 \quad , \tag{4.2b}$$

where $d\Omega$ is the element of solid angle $[d(\cos\theta)d\phi]$. Obviously, η_ℓ and therefore the cross section depends on the wave number k. This dependence is dropped in (4.1) and the following to simplify the notation. The focus for the remaining discussion will be the differential cross section $d\sigma/d\Omega$.

Rainbows are features of the differential cross section which, strictly speaking, are defined for the classical approximation to it. Although it is possible to begin a study of rainbows by starting directly with the classical expression for $d\sigma/d\Omega$, we prefer to arrive at the same results by following FORD and WHEELER [4.31], who first analysed the classical limit of (4.1). This approach has several important advantages because it gives insight into the nature and magnitude of errors that can be made by the purely classical approach.

To obtain the classical ($\hbar \to 0$) limit of (4.1), several distinct steps are taken. First, the summation over ℓ is replaced by an integral over ℓ. Classically, the magnitude of the orbital angular momentum ℓ can have arbitrary real values, thus the summation rigorously becomes an integration in the classical limit. Next, $P_\ell(\cos\theta)$ is replaced by the classical limit expressions which are obtained directly from the asymptotic expansions of $P_\ell(\cos\theta)$ for ℓ tending to infinity. They are

$$\sin\theta \gtrsim 1/\ell: \; P_\ell(\cos\theta) \underset{\ell\to\infty}{\sim} \left[\frac{1}{2}\left(\ell + \frac{1}{2}\right)\pi\sin\theta\right]^{-\frac{1}{2}} \sin\left[\left(\ell + \frac{1}{2}\right)\theta + \frac{\pi}{4}\right] \tag{4.3a}$$

$$\sin\theta \lesssim 1/\ell: \; P_\ell(\cos\theta) \underset{\ell\to\infty}{\sim} (\cos\theta)^\ell J_0\left[\left(\ell + \frac{1}{2}\right)\theta\right] \quad , \tag{4.3b}$$

where $J_0(x)$ is the regular Bessel function of zero order. Typically the rainbow scattering angle occurs for values of ℓ and θ such that (4.3a) is the appropriate classical limit expression for $P_\ell(\cos\theta)$. The classical limit of the phase shift η_ℓ is the well-known JWKB expression

$$\eta(\ell) = \lim_{R\to\infty} \left\{ \int_{R_0}^{R} \left[k^2 - \left(\ell + \frac{1}{2}\right)^2/R'^2 - U(R')\right]dR' - kR + \left(\ell + \frac{1}{2}\right)\frac{\pi}{2} \right\} \tag{4.4}$$

with R_0 the classical turning point and $U(R) = 2\mu V(R)/\hbar^2$. Inserting (4.3a,4) into (4.1) and replacing the summation by an integration, we have

$$f(\theta) \simeq -k^{-1}(2\pi\sin\theta)^{-\frac{1}{2}}(I_+ - I_-) \tag{4.5}$$

with

$$I_\pm = \int_0^\infty d\ell \left(\ell + \frac{1}{2}\right)^{\frac{1}{2}} \exp[i\phi_\pm(\ell)] \tag{4.6}$$

and

$$\phi_{\pm} = 2\eta(\ell) \pm (\ell + \frac{1}{2})\theta \pm \frac{\pi}{4} \quad . \tag{4.7}$$

The highly oscillatory integrand in (4.6) suggests that an approximate evaluation of the integral can be done by the method of stationary phase. Qualitatively, this method can answer the question as to whether $f(\theta)$ is large or small at a particular value of θ. If the phases $\phi_{\pm}$ are not stationary for a given scattering angle θ at any value of ℓ, then the oscillatory integrand will nearly cancel itself out and yield a small value for $f(\theta)$. Otherwise, if ϕ_+ and/or ϕ_- are stationary for at least one value of ℓ, then the integral will not cancel itself in the neighbourhood of the stationary point, although it will elsewhere, and $f(\theta)$ will not necessarily be small. The stationary condition for ϕ_+ is

$$\chi(\ell) \equiv 2\,\frac{d\eta(\ell)}{d\ell} = -\,\theta \tag{4.8a}$$

and

$$\chi(\ell) \equiv 2\,\frac{d\eta(\ell)}{d\ell} = \theta \tag{4.8b}$$

for ϕ_-. The quantity $\chi(\ell)$ can be shown to be the classical deflection function which can have any value between π and $-\infty$. For head on collisions, $\ell = 0$ and $\chi(\ell = 0)$ is always π for potentials which are repulsive at short range, as all atomic and molecular potentials are. Two "typical" deflection functions $\chi(\ell)$ are shown in Fig.4.2 for a potential which is attractive at long range. The collision energies are $E = 2\varepsilon$ and $E = 10\varepsilon$, respectively, with ε the potential well depth. For $\chi(\ell)$ greater than zero, the scattering is net repulsive and for $\chi(\ell)$ negative it is net attractive.

Figure 4.2 implies that ϕ_- *always* has one point of stationary phase ℓ_1 given by (4.8b). On the other hand, ϕ_+ has two points of stationary phase ℓ_2 and ℓ_3, given by (4.8a) if $\theta < |\chi_R|$ or none if $\theta > |\chi_R|$. The region around $\chi_R(\ell_R)$ is, therefore, an interesting one. In the vicinity of $\theta_R = |\chi_R|$ there is a range of stationary phase roots for nearly identical values of $\theta \simeq \theta_R$. This clearly implies that $f(\theta \simeq \theta_R)$ would be large and thus the differential cross section in the vicinity of θ_R would be intense. The scattering angle θ_R is called the rainbow angle and it is defined by

$$D(\ell) \equiv \frac{d\chi(\ell)}{d\ell} = 2\,\frac{d^2\eta(\ell)}{d\ell^2} = 0 \quad . \tag{4.9}$$

We can even get some qualitative idea of how intense the scattering will be at θ_R. Again from Fig.4.2 we see that if $\chi(\ell)$ is slowly varying in the vicinity of ℓ_R, then many values of ℓ will result in scattering angles nearly identical to θ_R. However, if $\chi(\ell)$ changes rapidly with ℓ away from $\chi_R(\ell_R)$, then fewer values of ℓ give rise to scattering angles approximately equal to θ_R. Thus we have two impor-

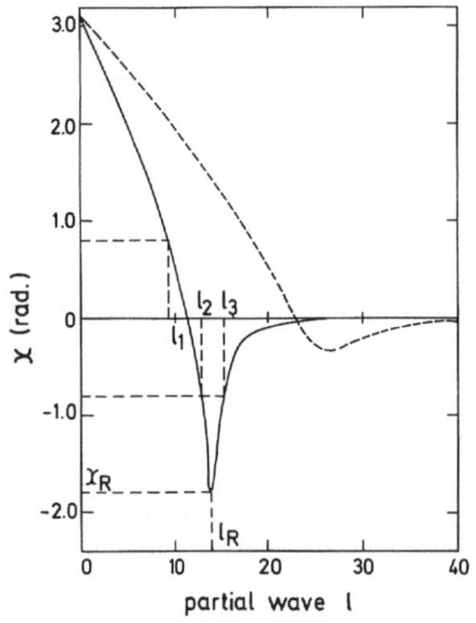

Fig.4.2. Schematic deflection function $\chi(\ell)$, (4.8), for two collision energies, $E = 2\varepsilon$ (——) and $E = 10\varepsilon$ (---) with ε the well depth. ℓ_1 is the root to (4.8b) and ℓ_2 and ℓ_3 are the roots to (4.8a) for $\theta = 0.8$ radiant. $\theta_R = |\chi_R|$ is the classical rainbow angle and ℓ_R is defined in (4.9)

tant qualitative conclusions which can be stated in the following form: (a) intense scattering occurs at θ_R where $\theta_R = |\chi_R(\ell_R)|$ and ℓ_R is the solution to (4.9) and (b) the intensity at θ_R is proportional to the broadness of $\chi(\ell)$ in the vicinity of ℓ_R.

Let us now see how these qualitative results can be made quantitative (or semi-quantitative, at least). First we evaluate the integrals in (4.6) by the usual method of stationary phase [4.32]. The result is

$$I_+ = (2\pi i)^{\frac{1}{2}} \sum_{\nu=2,3} \left[\frac{(\ell_\nu + \frac{1}{2})}{D(\ell_\nu)}\right]^{\frac{1}{2}} \exp i\phi_+(\ell_\nu) \tag{4.10a}$$

for I_+ with ℓ_2 and ℓ_3 defined in (4.8a) and

$$I_- = (2\pi i)^{\frac{1}{2}} \left[\frac{(\ell_1 + \frac{1}{2})}{D(\ell_1)}\right]^{\frac{1}{2}} \exp[i\phi_-(\ell_1)] \tag{4.10b}$$

for I_- with ℓ_1 defined in (4.8b) and $D(\ell)$ is given in (4.9). Equation (4.10) is valid for real and complex stationary phase points. While ℓ_1 is always real, ℓ_2 and ℓ_3 become complex, i.e., classically forbidden if $\theta > \theta_R$.

As already noted from Fig.4.2, $\chi(\ell)$ is stationary at ℓ_R and we argued that at the corresponding angle θ_R, intense scattering would be observed. The potential acts like a lens focussing the flux at the scattering angle θ_R. This effect appears according to (4.10a) as a singularity in $f(\theta)$ and $d\sigma(\theta)/d\Omega$ because D becomes zero at the rainbow angle. Clearly something is wrong with the mathematics at this value of θ; the idea, however, is right. The basic problem is that $\phi_+(\ell)$ is stationary to first *and* second order at ℓ_R and so the usual expansion to second order is inadequate. Expanding to third order is necessary about ℓ_R and yields for $d\sigma/d\Omega$ in the vicinity of and including θ_R [4.31]:

$$\frac{d\sigma}{d\Omega}(\theta) = \frac{2\pi(\ell_R + \frac{1}{2})}{k^2 \sin\theta} q^{-2/3} \mathrm{Ai}^2(x) \quad , \tag{4.11}$$

where

$$q = \frac{1}{2} \frac{d^2\chi(\ell)}{d\ell^2}\bigg|_{\ell=\ell_R} \tag{4.12a}$$

$$x = q^{-1/3}(\theta - \theta_R) \tag{4.12b}$$

and Ai(x) is the regular Airy function. In (4.11) the contribution of I_- has been neglected. Examining (4.11) for $\theta \simeq \theta_R$ shows a "rainbow maximum" in $d\sigma/d\Omega$ not an infinity of course, and a width which scales like $q^{-1/3}$. Thus, if $d^2\chi(\ell)d\ell^2$ is small at ℓ_R, $q^{-1/3}$ is large and $d\sigma/d\Omega$ changes rapidly from its maximum value. This leads to a narrow, intense rainbow feature, in agreement with our earlier qualitative conclusion. We also see from (4.11) that the actual maximum in $d\sigma/d\Omega$ will be shifted somewhat from θ_R to a smaller angle θ_{max} because $\mathrm{Ai}^2(x)$ has a maximum for $x = -1.0188$, not for $x = 0$. Thus, (4.12b) leads to a simple relationship between θ_R and θ_{max} which is

$$\theta_{max} = \theta_R - 1.0188\, q^{1/3} \quad . \tag{4.13}$$

Since q is positive, the quantal rainbow maximum is always shifted to smaller angles away from the classical singularity. The shift in the rainbow angle increases as q increases. In summary, if $d^2\chi(\ell)/d\ell^2$ is large the rainbow feature will be broad and shifted considerably from θ_R.

Since the three contributions [ℓ_1, ℓ_2 and ℓ_3 in (4.10)] to the classical limit scattering amplitude have, in general, different phases, the differential cross section shows pronounced interference patterns [4.30]. The interference of ℓ_2 and ℓ_3 in Fig.4.2 yields oscillations with large angular separations (supernumerary rainbows), whereas the interference of ℓ_1 with ℓ_2 and ℓ_1 with ℓ_3 yields superimposed fast oscillations. These oscillations die out for large scattering angles well separated from the rainbow angle θ_R.

Let us return to (4.6) to take another point of view of rainbow scattering. We have already said that the integrand will approximately cancel itself if $\phi_\pm(\ell)$ has no stationary points and will give a very small $d\sigma/d\Omega$. If $\phi_\pm(\ell)$ is stationary, then most of the integral can be evaluated by considering the vicinity of the stationary phase point. At an ordinary stationary point, $\phi_\pm(\ell)$ is stationary only to first order in ℓ. Obviously, if $\phi_\pm(\ell)$ is stationary to first and second-order variations of ℓ, the resulting integral will be even larger than it would be for an ordinary stationary point. That special stationary point is the rainbow ℓ, ℓ_R. There is finally one other point of view we wish to consider to characterize rainbow scattering. The rainbow angle is the boundary between the regions with three and one stationary phase points for $f(\theta)$. For $\theta < \theta_R$ there are three stationary phase contributions to $f(\theta)$ but for $\theta > \theta_R$ there is only one stationary phase contribution to $f(\theta)$. The two roots to (4.8a) are limited to scattering angles less than θ_R. As a result $f(\theta)$ and $d\sigma/d\Omega$ are expected to be small for $\theta > \theta_R$ with the transition occuring at $\theta = \theta_R$.

In summary, there are several equivalent ways to analyse and identify rainbow scattering in atom-atom, isotropic potential scattering, all based on the stationary phase method of evaluating the classical limit of $f(\theta)$. These various methods each provide a slightly different point of view of rainbow scattering and together they give clear physical insight to the phenomenon.

As we shall see in the next section, the rainbow analysis for rotationally inelastic scattering is more complicated, with the several points of view just discussed for central field scattering not exactly giving equivalent descriptions of the rainbow effect. However, by considering all of them we hope to gain maximum physical insight into these new kinds of rainbow features.

4.3 Classical Limit of the IOS Approximation

4.3.1 Basic IOS Formulas

The IOS approximation has been reviewed recently [4.20] and therefore we give only the basic equations necessary for the present purpose. The vibrational degree of freedom is neglected in what follows, i.e., the rigid-rotor model is employed. Identifying the average orbital angular momentum as the final orbital angular momentum yields the simplest expression for the scattering amplitude for a $j_1m_1 \to j_2m_2$ transition, i.e.,

$$f(j_1m_1 \to j_2m_2|\theta) = \frac{i(-1)^{j_1+j_2}}{2(k_{j_1}k_{j_2})^{1/2}}\,\delta_{m_1m_2} \sum_\ell (2\ell+1)T_m^\ell(j_1 \to j_2)P_\ell(\cos\theta) \quad , \tag{4.14}$$

where $m = \min(m_1,m_2)$ and

$$T_m^\ell(j_1 \to j_2) = 2\pi \int_0^\pi d\gamma\, \sin\gamma\, Y_{j_1m}(\gamma,0)\{1 - \exp[2i\eta_\ell(\gamma)]\}Y_{j_2m}(\gamma,0) \quad . \tag{4.15}$$

In (4.14), k_j is the wave number for state j, i.e.,

$$k_j = \{2\mu[E + B_{rot}j_1(j_1+1) - B_{rot}j(j+1)]\}^{1/2}/\hbar \quad . \tag{4.16}$$

The elastic and inelastic scattering amplitudes in (4.1,14) differ mainly in the definition of the T-matrix. In (4.15), Y_{j_1m} and Y_{j_2m} are spherical harmonics which stand for the initial and final rotor states, respectively. The phase shift $\eta_\ell(\gamma)$ depends, through the potential, on the orientation angle γ.

The "selection rule" $\Delta m = 0$ in (4.14), which results from the special choice of the average orbital angular momentum, is irrelevant, when only degeneracy-averaged cross sections

$$\frac{d\sigma}{d\Omega}(j_1 \to j_2|\theta) = \frac{1}{2j_1+1}\frac{k_{j_2}}{k_{j_1}}\sum_{m_1m_2} |f(j_1m_1 \to j_2m_2|\theta)|^2 \tag{4.17}$$

are considered [4.20]. We shall restrict our discussion to these. In addition, we confine the analysis of the classical limit of the IOS approximation to scattering out of the ground state ($j_1 = m_1 = 0$) because any $j_1 \to j_2$ cross section can be obtained from the $0 \to j_2$ cross sections according to [4.20]

$$\frac{d\sigma}{d\Omega}(j_1 \to j_2|\theta) = \sum_j C^2(j_1jj_2|000)\,\frac{d\sigma}{d\Omega}(0 \to j|\theta) \quad , \tag{4.18}$$

where $C(j_1jj_2|000)$ is a Clebsch-Gordan coefficient.

4.3.2 Classical Limit of (4.14) for j = 0

For initially nonrotating molecules ($j_1 = m_1 = 0$), the sum in (4.17) reduces to a single term with $m_1 = m_2 = 0$ and the scattering amplitude is

$$f(0 \to j_2|\theta) = \frac{i}{4}(-1)^{j_2}(k_0k_{j_2})^{-\frac{1}{2}}(2j_2+1)^{\frac{1}{2}}$$

$$\sum_\ell (2\ell+1)P_\ell(\cos\theta)\int_0^\pi d\gamma\,\sin\gamma$$

$$P_{j_1=0}(\cos\gamma)\{1 - \exp[2i\eta_\ell(\gamma)]\}P_{j_2}(\cos\gamma) \quad . \tag{4.19}$$

To obtain the classical limit of (4.19), we would simply replace all the Legendre polynomials by their well-known classical limit expression (4.3a), convert the summation over ℓ into an integral over ℓ (treated as a continuous variable, of course) and replace the phase shift $\eta_\ell(\gamma)$ by its JWKB limit (4.4). However, the replacement of $P_{j=0}(\cos\gamma) = 1$ by its classical limit, while strictly correct, has been shown [4.26] to lead to large errors in the resulting cross sections compared to those where the quantum P_0 and the classical limit of P_{j_2} are used. Thus, we shall consider the latter hybrid approach and continue, following KORSCH and SCHINKE [4.27], by rewriting (4.19) as

$$f(0 \to j_2|\theta) = -\frac{(-1)^{j_2}}{2\pi}(2k_0k_{j_2}\sin\theta)^{-\frac{1}{2}}(I_+^+ - I_-^- + iI_-^+ + iI_+^-) \quad , \tag{4.20}$$

where

$$I_\pm^\pm = \int_0^\infty d\ell \int_0^\pi d\gamma[(\ell+\tfrac{1}{2})\sin\gamma]^{\frac{1}{2}}\exp[i\phi_\pm^\pm(\ell,\gamma)] \tag{4.21}$$

$$\phi_{\pm}^{\pm} = 2\eta(\ell,\gamma) \pm \theta(\ell + \tfrac{1}{2}) \pm (j_2 + \tfrac{1}{2})\gamma \quad . \tag{4.22}$$

The upper and lower indices of I refer to the sign of $(j_2 + 1/2)$ and θ in the phase, respectively. Equations (4.20-22) are the extension of (4.5-7) for rotationally inelastic scattering within the IOS approximation. To make contact with classical mechanics we evaluate the two-dimensional integral by the stationary phase method [4.32] to obtain

$$I_{\pm}^{\pm} = 2\pi i \sum_{\nu} \left[\frac{(\ell_\nu + \frac{1}{2})\sin\gamma_\nu}{D(\ell_\nu,\gamma_\nu)}\right]^{\frac{1}{2}} \exp[i\phi_{\pm}^{\pm}(\ell_\nu,\gamma_\nu)] \quad , \tag{4.23}$$

where

$$D(\ell,\gamma) = \det\begin{pmatrix} 2\,\dfrac{\partial^2\eta(\ell,\gamma)}{\partial\ell^2} & 2\,\dfrac{\partial^2\eta(\ell,\gamma)}{\partial\ell\partial\gamma} \\[2ex] 2\,\dfrac{\partial^2\eta(\ell,\gamma)}{\partial\gamma\partial\ell} & 2\,\dfrac{\partial^2\eta(\ell,\gamma)}{\partial\gamma^2} \end{pmatrix} \quad . \tag{4.24}$$

The summation in (4.23) is over all roots (ℓ_ν,γ_ν) to the coupled stationary phase equations

$$\chi(\ell,\gamma) \equiv 2\,\frac{\partial\eta(\ell,\gamma)}{\partial\ell} = \mp\,\theta \tag{4.25}$$

$$J(\ell,\gamma) \equiv 2\,\frac{\partial\eta(\ell,\gamma)}{\partial\gamma} = \mp\,(j_2 + \tfrac{1}{2}) \tag{4.26}$$

which relate the final scattering conditions, i.e., θ and j_2, to the classical deflection functions $\chi(\ell,\gamma)$ and $J(\ell,\gamma)$, respectively. Recall that χ was given earlier but for isotropic potential scattering. The new "deflection function" $J(\ell,\gamma)$ is the (approximate) "final action trajectory function" of classical S-matrix theory [4.23,24] (Sect.4.3.5). Since $J(\ell,\gamma)$ describes the excitation of the rotor, we shall sometimes call it the excitation function. Also note that (4.23) is the extension of (4.10) to two dimensions.

Rainbow-like singularities in the classical limit of the IOS cross section will occur whenever the determinant

$$\begin{aligned} D(\ell,\gamma) &= 4[(\partial^2\eta/\partial\ell^2)_\gamma(\partial^2\eta/\partial\gamma^2)_\ell - (\partial^2\eta/\partial\ell\partial\gamma)^2] \\ &= (\partial\chi/\partial\ell)_\gamma(\partial J/\partial\gamma)_\ell - (\partial\chi/\partial\gamma)_\ell(\partial J/\partial\ell)_\gamma \end{aligned} \tag{4.27}$$

becomes zero for a particular set (ℓ_R,γ_R) which in turn defines a set (θ_R,j_R) through the stationary phase equations (4.25,26), respectively. Note that this determinant reduces to that in (4.9) of the previous section if $V(R,\gamma)$ is independent of γ, as it is in atom-atom potential scattering.

Suppose $D(\ell,\gamma)$ can be approximated by its diagonal elements, i.e.,

$$D(\ell,\gamma) = (\partial\chi/\partial\ell)_\gamma(\partial J/\partial\gamma)_\ell \quad . \tag{4.28}$$

Then it would be possible to classify rainbows according to whether $(\partial\chi/\partial\ell)_\gamma$ or $(\partial J/\partial\gamma)_\ell$ vanishes. The first rainbow would then correspond to the usual rainbow seen in atom-atom potential scattering and the second rainbow would represent a new kind which occurs for rotationally inelastic scattering. It is termed a *rotational rainbow*. This approximation to (4.27) is a good one if the two inequalities

$$(\partial J/\partial\gamma)_\ell \gg (\partial J/\partial\ell)_\gamma; \qquad (\partial\chi/\partial\ell)_\gamma \gg (\partial\chi/\partial\gamma)_\ell$$

are satisfied. It has been used in a number of early studies of rotational rainbows [4.21,25,27-29]. Guided by experience we conclude that the above inequalities are usually fulfilled for impulsive collisions, i.e., large wave numbers and dominantly repulsive potentials. For collision systems with substantial γ-dependent potential minima at a range where rotational inelasticity is still strong, the above assumptions may not hold. Recently the accuracy of (4.28) has been explicitly verified [4.16,33] (see also Sect.4.5.1) and as we shall see, conclusions based on it are in excellent qualitative agreement with experiment. This fact, coupled with the insight obtained by making this approximation, justifies, in our opinion, the use of (4.28).

Physically the argument justifying the above classification of rainbows is best seen from an example. Consider a fixed j_2 and suppose there is a range of values of ℓ where there are several roots to (4.26). As ℓ increases suppose two of these roots begin to coalesce so that at a particular value of ℓ and γ (call them ℓ_R and γ_R), they have coalesced. (The situation then is quite analogous to the one discussed in Sect.4.2 for isotropic scattering at the rainbow scattering angle.) Thus, a range of values of γ in the vicinity of γ_R (for $\ell = \ell_R$) leads to the same value of J, i.e., $j_2 + 1/2$. Provided $\chi(\ell_R,\gamma)$ does not vary with respect to γ in the vicinity of γ_R more rapidly than at other values of γ, we would predict intense scattering for the $0 \to j_2$ transition at the scattering angle $\theta = \chi(\ell_R,\gamma_R)$. A similar analysis can be made for χ as a function of ℓ(just as in Sect.4.2) with the analogous result, provided that at the value of ℓ where roots to (4.26) begin coalescing, $J(\ell,\gamma)$ is changing slowly with respect to ℓ. The two inequalities (above) upon which (4.28) is based are completely consistent with the assumption we have made in this physically based argument (they are, in fact, stronger assumptions than the ones just discussed).

For the collision systems we will review in this chapter, the collision energies are always much greater than the well depth of the interaction potential and the above inequalities [just after (4.28)] are satisfied, in general. To simplify the discussion, we assume in the following dominantly repulsive potentials. However, we note that the main conclusions are also valid for systems with weak attraction provided the scattering angle is large.

4.3.3 Uniform Approximation

For a purely repulsive potential surface, (4.25) has exactly one real solution for the plus but none for the minus sign, and only the $I_-^{\pm}$ integrals in (4.21) contribute to the scattering amplitude. Then (4.25) establishes for a fixed orientation γ a one-to-one correspondence between partial wave ℓ and scattering angle θ. To locate the solutions of the second stationary phase condition (4.26), we plot in Fig.4.3a a representative example of $\eta(\ell,\gamma)$ for a heteronuclear molecule versus γ for a fixed ℓ. Since the respective potential surface is purely repulsive, the phase shift is negative for all ℓ and γ. The scattering for collinear orientations ($\gamma = 0^{\circ}$ and 180°) is usually more repulsive than for noncollinear ones and so the phase shift has a maximum around 90°. Because of the asymmetry of the target molecule, $\eta(\ell,\gamma)$ is different for $\gamma = 0^{\circ}$ and $\gamma = 180^{\circ}$. The corresponding final action function $J(\ell,\gamma)$ is shown in Fig.4.3b. The two branches of $J(\ell,\gamma)$ correspond to scattering from the two different ends of the heteronuclear molecule. In particular, for a given j_2 [$j_2 = 30$ in Fig.4.3b], two solutions for each branch exist which are labeled γ_1, γ_2 and γ_3, γ_4, respectively. The orientations γ_1 and γ_2 contribute to the integral I_-^- and γ_3 and γ_4 contribute to I_-^+ in (4.21).

The stationary phase equations (4.25,26) define contours $\ell(\gamma;\theta)$ for fixed scattering angle θ and $\gamma(\ell;j_2)$ for fixed j_2, respectively, as shown in Fig.4.4. Because both equations must be satisfied simultaneously for given values of θ and j_2, only the intersections of the corresponding contours (ℓ_ν,γ_ν), $(\nu = 1,\ldots,4)$ contribute to the cross section. The rainbow occurs for (θ_R,j_R) when the corresponding contours $\ell(\gamma;\theta_R)$ and $\gamma(\ell;j_R)$ are tangent at (ℓ_R,γ_R) and the determinant (4.27) is zero. An example can be seen in Fig.4.4 for $\theta = 30^{\circ}$ and $j_2 = 20$ ($\ell_R \simeq 390$; $\gamma_R \simeq 50^{\circ}$). As in the case of isotropic potential scattering the simple formula (4.23) breaks down at a rainbow point and the cross section rises to infinity. If two contours do not intersect for real values of ℓ and γ, they can be analytically continued into the complex plane and complex intersections can be searched for there. In general, there are four real, or two real and two complex, or four complex roots. Examples for these three cases are seen in Fig.4.4 for ($\theta = 150^{\circ}$, $j_2 = 30$), ($\theta = 90^{\circ}$, $j_2 = 30$) and ($\theta = 30^{\circ}$, $j_2 = 30$), respectively. The scattering amplitude corresponding to the complex roots will be expected to be small, of course.

As observed in Fig.4.4 only two points $\nu = 1$ and $\nu = 2$ *or* $\nu = 3$ and $\nu = 4$ coalesce at the same time and, therefore, it is possible to derive cross section formulas which are uniformly valid for *all* transitions and *all* angles [4.32]. The uniform semiclassical cross section is finally given by [4.27]

$$\frac{d\sigma}{d\Omega}(0 \to j_2|\theta) = \frac{\pi}{2k_0^2 \sin\theta} |A_{21} - iA_{34}| \tag{4.29a}$$

$$= \frac{\pi}{2k_0^2 \sin\theta}\left(|A_{21}|^2 + |A_{34}|^2 + 2\,\mathrm{Re}\{iA_{21}A_{34}^*\}\right) \tag{4.29b}$$

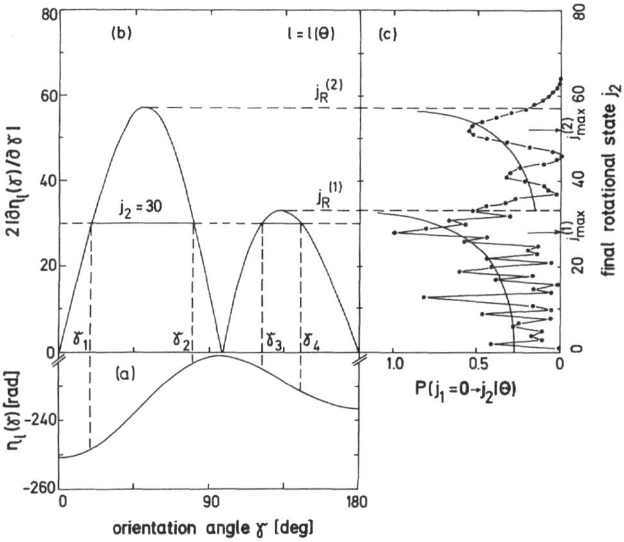

Fig.4.3. (a) JWKB phase shift $\eta(\ell,\gamma)$ versus orientation angle γ for fixed partial wave ℓ. The potential energy surface is that of SCHINKE [4.29], representing roughly K - CO scattering, and the collision energy is E = 1.24 eV. (b) Excitation function $J(\ell,\gamma)$ (4.26), versus γ at fixed partial wave ℓ. $\gamma_1,\ldots,\gamma_4$ are the roots to (4.26) for fixed ℓ and $j_2 = 30$. γ_1 and γ_2 are the roots for the plus sign, γ_3 and γ_4 are the roots for the minus sign in (4.26). $j_R^{(1)}$ and $j_R^{(2)}$ are the classical rainbow states for this particular value of ℓ. (c) IOS differential cross sections (normalized to one) versus final rotational state j_2 for fixed scattering angle θ. A unique relation $\ell = \ell(\theta)$ is assumed which is independent of γ and j_2. $j_{max}^{(1)} < j_R^{(1)}$ and $j_{max}^{(2)} < j_R^{(2)}$ are the quantal rainbow states. The smooth line represents the classical cross section

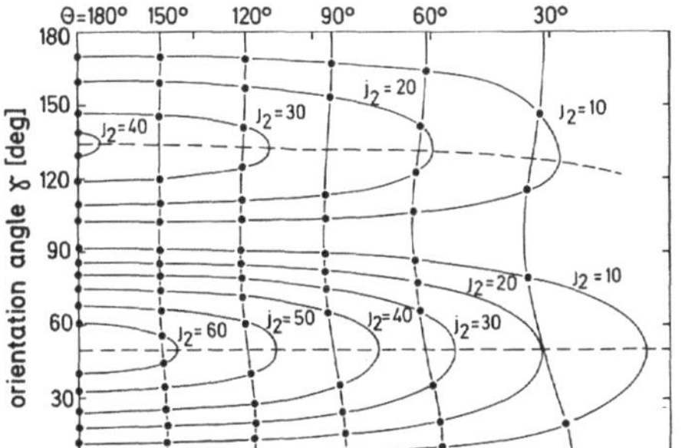

Fig.4.4. Contours $\ell = \ell(\gamma;\theta)$ for fixed scattering angle θ and $\gamma = \gamma(\ell;j_2)$ for fixed final state j_2 as obtained from (4.25,26), respectively. Potential and scattering parameters are the same as in Fig.4.3, i.e., K-CO. The points (● ●) represent the stationary phase points (ℓ_ν,γ_ν). The dashed line represents the rainbow curve, i.e., $D(\ell_R,\gamma_R)=0$ in (4.27)

with

$$A_{\mu\nu} = \exp[\tfrac{i}{2}(\phi_\mu + \phi_\nu)][(F_\mu + F_\nu)\xi_{\mu\nu}^{1/4}Ai(-\xi_{\mu\nu}) + i(F_\mu - F_\nu)\xi_{\mu\nu}^{-1/4}Ai'(-\xi_{\mu\nu})] \quad . \tag{4.30}$$

The phases are defined by

$$\phi_\nu = 2\eta(\ell_\nu,\gamma_\nu) - \theta(\ell_\nu + \tfrac{1}{2}) \mp (j_2 + \tfrac{1}{2})\gamma_\nu \quad , \tag{4.31}$$

where the -/+ sign for the last term corresponds to the two sets of roots $\nu = 1,2$ and $\nu = 3,4$, respectively. The classical probability factors are given by

$$F_\nu = \left(\frac{(\ell_\nu + \frac{1}{2})\sin\gamma_\nu}{|D(\ell_\nu,\gamma_\nu)|}\right)^{\frac{1}{2}} \tag{4.32}$$

and the argument of the Airy function Ai and its derivative Ai' is defined by

$$\xi_{\mu\nu} = \left[\frac{3}{4}\,(\phi_\mu - \phi_\nu)\right]^{2/3} \quad . \tag{4.33}$$

The cross section formulas become considerably simpler for homonuclear molecules. Because then $\eta(\ell,\gamma)$ is symmetric about $\pi/2$, γ_3 and γ_4 are related to γ_1 and γ_2 according to

$$(\ell_3,\gamma_3) = (\ell_2,\pi - \gamma_2) \tag{4.34a}$$

$$(\ell_4,\gamma_4) = (\ell_1,\pi - \gamma_1) \tag{4.34b}$$

and so

$$A_{34} = \exp\left[i(j_2 + \tfrac{1}{2})\pi\right]A_{21} \quad . \tag{4.35}$$

Then the cross section reduces to

$$\frac{d\sigma}{d\Omega}\,(0 \to j_2|\theta) = \frac{\pi}{k_0^2\,\sin\theta}\,|A_{21}|^2\left[1 + \sin(j_2 + \tfrac{1}{2})\pi\right] \tag{4.36}$$

which is zero for odd j_2 as it should be. Thus, the selection rule that Δj must be even for homonuclear molecules follows directly from interference of amplitudes as previously and elegantly demonstrated by classical S-matrix theory [4.34]. The cross section for homonuclear molecules can be explicitly written as

$$\frac{d\sigma}{d\Omega}\,(0 \to j_2|\theta) = \frac{2\pi}{k_0^2\,\sin\theta}\,[(F_2 + F_1)^2\xi^{\frac{1}{2}}\mathrm{Ai}^2(-\xi) + (F_2 - F_1)^2\xi^{-\frac{1}{2}}\mathrm{Ai}'^2(-\xi)] \tag{4.37}$$

for classically allowed transitions ($\xi = \xi_{21} = \xi_{34}$) and

$$\frac{d\sigma}{d\Omega}\,(0 \to j_2|\theta) = \frac{4\pi}{k_0^2\,\sin\theta}\,|F|^2[(1 + \sin\beta)|\xi|^{\frac{1}{2}}\mathrm{Ai}^2(|\xi|) + (1 - \sin\beta)|\xi|^{-\frac{1}{2}}\mathrm{Ai}'^2(|\xi|)] \tag{4.38}$$

for classically forbidden transitions when (ℓ_1,γ_1) and (ℓ_2,γ_2) are complex conjugate $(|F| = |F_1| = |F_2|$ and $\beta = 2 \arg F_2 = \pi - 2 \arg F_1)$. For classically allowed transitions, well separated from the rainbow singularity, (4.37) becomes [4.27]

$$\frac{d\sigma}{d\Omega}(0 \to j_2|\theta) = \frac{2}{k_0^2 \sin\theta}\left[F_1^2 + F_2^2 + 2F_1F_2 \sin(\phi_1 - \phi_2)\right] . \tag{4.39a}$$

The corresponding expression for classically forbidden transitions is [4.27]

$$\frac{d\sigma}{d\Omega}(0 \to j_2|\theta) = \frac{2}{k_0^2 \sin\theta} |F|^2 \exp(-2 \operatorname{Im}\{\phi\}) . \tag{4.39b}$$

A classical approximation to (4.39a) is obtained by neglecting the interference term. It is

$$\frac{d\sigma}{d\Omega}(0 \to j_2|\theta) = \frac{2}{k_0^2 \sin\theta}(F_1^2 + F_2^2) \tag{4.40}$$

and is only valid for even j_2. Because the phase information is discarded in (4.40), the classical cross section for odd j_2 would be nonzero. It is obviously zero for classically forbidden transitions.

This semiclassical treatment of the quantal IOS approximation is in several respects reminiscent of classical S-matrix theory in the j_z-conserving approximation [4.34]. In Sect.4.3.5 we shall explicitly point out the similarities and differences between the two theories.

In [4.27] a detailed comparison between uniform semiclassical and quantal IOS cross sections is performed for homonuclear as well as for heteronuclear molecules and excellent agreement is obtained. A similar study will be discussed below in Sect.4.4.1. Since the exact quantal IOS approximation is extremely simple to implement even if transitions to very high rotational states are considered, the uniform semiclassical IOS approximation is not advantageous in numerical studies. Its great asset is its power to interpret the qualitative behavior of rotationally inelastic cross sections.

4.3.4 Qualitative Behavior of Cross Sections

From Fig.4.4 we observe that especially for large scattering angles the contour $\ell(\gamma;\theta)$ depends only weakly on γ and thus the stationary phase partial wave $\ell_\nu(\nu = 1,\dots,4)$ is well approximated by a constant for given θ. This defines a one-to-one mapping between ℓ and θ *independent of* γ and j_2. Therefore, it is possible to discuss the j_2-distribution at a fixed scattering angle θ in terms of a single excitation function $J(\ell(\theta),\gamma)$. This has been exploited in earlier studies of rotational rainbow scattering [4.21,25,29] and it is illustrated in Figs.4.3b,c. Obviously, the weak γ-dependence of $\ell(\gamma;\theta)$ is equivalent to the second inequality from above just after (4.28).

For small j_2 there are four real stationary phase angles γ_ν and the corresponding amplitudes will strongly interfere with each other. The first two terms in (4.29b) describe separate intereference between the two roots of each branch. Because the interfering contributions belong to the same branch this type of interference is expected to yield oscillations in the cross section with a wide j_2 spacing. In analogy with elastic scattering from isotropic potentials we shall call the resulting secondary maxima *supernumerary rotational rainbows*. The cross term in (4.29b) describes interference between the two branches and gives rise to rapid oscillations. For small j_2 the supernumerary oscillations of both branches, which are due to the sin-term in (4.39a), and the rapid oscillations overlap and, thus, the cross section does not show particularly simple behavior. As j_2 increases, γ_3 and γ_4 coalesce and finally for $j_2 = j_R^{(1)}$ the first classical rainbow occurs. Rigorously, $j_R^{(1)}$ marks the boundary for the second branch where values of j_2 greater than $j_R^{(1)}$ are "classically forbidden", i.e., there are no real contributions to I_-^+ in (4.21). The quantum (or uniform semiclassical) cross section shown in Fig.4.3c has an oscillatory maximum at a value of $j_2 = j_{max}^{(1)}$ which is somewhat less than $j_R^{(1)}$. The oscillations are expected because of the four roots which contribute to the $j_2 = j_R^{(1)}$ scattering amplitude, two have coalesced but the other two have not. Phases associated with the noncoalescent roots interfere with each other and with the phases corresponding to the coalescent roots. That $j_{max}^{(1)}$ shown is less than $j_R^{(1)}$ can be explained by the uniform analysis [see (4.41) below]. For $j_2 > j_R^{(1)}$, γ_3 and γ_4 become complex and $|A_{34}|^2$ decreases rapidly because the argument of the Airy function is positive. With increasing j_2 the oscillations become more regular because the contribution from the second branch has died off and only the amplitudes for γ_1 and γ_2 give rise to interference. At $j_R^{(2)}$ the second classical rainbow occurs. Again, the quantum or uniform semiclassical cross section has a maximum at $j_{max}^{(2)} < j_R^{(2)}$, which is expected to be smooth because there is only one non-negligible amplitude A_{21} contributing to (4.29). Transitions with $j_2 > j_R^{(2)}$ are classically-forbidden at this scattering angle and the cross section smoothly decreases to zero.

The two branches in Fig.4.3b become equivalent for homonuclear molecules and the two rainbow states become identical $[j_R^{(1)} = j_R^{(2)} = j_R]$. Simultaneously, the oscillation structure in the differential cross section becomes simpler because only two orientation angles contribute to a given transition. Thus, supernumerary rotational rainbows are expected to be clearly visible for homonuclear molecules. They stem from the sin-term in (4.39) and are a direct consequence of the quantum mechanical superposition principle. Therefore, they cannot be discribed by any classical approximation, for example, (4.40). The cross section for odd j_2 transitions is zero.

The uniform Airy analysis leads to a simple relationship between j_R where the classical cross section is infinite, and j_{max} where the quantum cross section has a rainbow maximum. For simplicity consider the case of a homonuclear rigid-rotor.

Expanding $J(\ell,\gamma)$ for fixed ℓ about the rainbow orientation angle γ_R, one approximately obtains [4.27]

$$j_{max} = j_R - 1.0188\, q^{1/3} \tag{4.41}$$

with

$$q = -\frac{1}{2}\frac{\partial^2 J(\ell,\gamma)}{\partial^2\gamma^2}\bigg|_{\gamma=\gamma_R} \quad . \tag{4.42}$$

As seen in Fig.4.3b, $\partial^2 J/\partial\gamma^2$ is negative for the first branch and so q is positive. It follows that the quantum rainbow maximum is always shifted to smaller j_2 values away from the classical singularity, i.e., $j_{max} < j_R$. The shift scales with the second derivative of the excitation function $J(\ell,\gamma)$ with respect to γ calculated at the rainbow angle γ_R. Similar relations hold for both rotational rainbow maxima in the case of heteronuclear molecules. The supernumerary maxima which occur in the classically allowed j_2 region ($j_2 < j_{max}$) correspond to the additional maxima of the Airy function at $x = -3.2482, -4.9201$, etc. [Ref.4.35, Table 10.13]. The similarity of this discussion to the analogous, well-known uniform analysis of atom-atom isotropic scattering should be noted, see (4.13).

Consider now how rotational rainbows appear in the differential cross section for a fixed j_2 as a function of θ. For simplicity we again restrict the discussion to homonuclear molecules. For the general case of heteronuclear see [4.27]. The equation $D(\ell,\gamma) = 0$ defines the rainbow points (θ_R, j_R), i.e., the classical rainbow curve $j_R(\theta)$ which for fixed angle θ locates the rainbow singularity in the j_2-distribution. The inverse function $\theta_R(j_2)$ (provided it exists) gives the scattering angle where the classical differential cross section diverges for a given j_2. A typical example of $j_R(\theta)$ is given in Fig.4.5a. As seen, j_R increases with θ or equivalently θ_R is an increasing function of j_2. Such behavior is surmised by inspection of Fig.4.4 and is characteristic for dominantly repulsive potential surfaces. The maximum of the excitation function which determines j_R, decreases with increasing ℓ, when the rotational coupling becomes weaker. This has been observed in various computational studies concerning model [4.29] or realistic [4.16,25,27] systems. A particular example will be discussed in the next section (Fig.4.6). For repulsive potentials θ is also a monotonically decreasing function of ℓ (almost independent of γ) such that j_R increases with θ (and vice versa). More complicated behavior of the rainbow curve $j_R(\theta)$ is expected for potentials with an appreciably deep well especially if the anisotropy is strong in the attractive region.

In Fig.4.5a the curve $j_{max}(\theta)$ represents the quantum or uniform semiclassical rainbow curve. Its shift from $j_R(\theta)$ is approximately given by (4.41). At any angle θ values of j_2 less than $j_R(\theta)$ are classically allowed and those above it are classically forbidden. Equivalently, for a fixed j_2, the angular region between zero and $\theta_R(j_2)$ is classically forbidden and that beyond $\theta_R(j_2)$ is classically

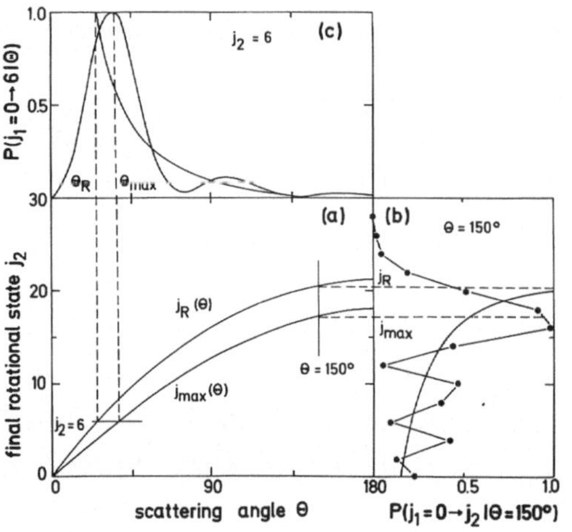

Fig.4.5. (a) Classical and quantal rainbow curves $j_R(\theta)$ and $j_{max}(\theta)$, respectively, for He-Na_2. The potential is that of [4.51] and E = 100 meV. (b) IOS differential cross sections (normalized to one) versus j_2 for θ = 150°. j_R and j_{max} are the classical and quantal rotational rainbow states at this angle. (c) IOS differential cross sections (normalized to one) versus θ for j_2 = 6. θ_R and θ_{max} are the classical and quantal rotational rainbow angles for this transition. The smooth curves in (b) and (c) represent the classical cross section (4.40)

allowed. Cross sections can be represented as j_2-distributions at fixed scattering angles or as θ-distributions for fixed j_2 transitions, as shown in Figs.4.4b,c, respectively. The first case has already been discussed in Fig.4.3c. In the latter case the cross section rises steeply out of the classically forbidden region in the forward direction and reaches a maximum at $\theta_{max}(j_2)$ which by inspection of Fig. 4.5 is greater than $\theta_R(j_2)$. Then it declines into the classically allowed angular region and shows pronounced supernumerary rainbow maxima. The rainbow structure in both representations (Figs.4.5b,c) is totally equivalent [4.21,27]. Because the rainbow curve levels out for large scattering angles, the shift $\theta_{max}(j_2) - \theta_R(j_2)$ increases considerably with j_2. Thus, a purely classical description becomes gradually less accurate in locating the rainbow maxima $\theta_{max}(j_2)$ for large j_2 and wide scattering angles [4.33]. The classical cross section (4.40) which is also plotted in Fig.4.5 monotonically falls off into the classically allowed j_2 or θ region and averages over the quantum oscillations. It is exactly zero in the classically forbidden region. The experimental angular distributions shown in Sect.4.1.1 (Fig.4.1) behave exactly as predicted in Fig.4.5c. Incidentally, we note that θ-distributions are not suitable to most clearly display the supernumerary rainbow maxima. Because of the $1/\sin\theta$ factor in all cross section expressions, the intensity of the supernumeraries decreases rapidly as θ increases. Thus, only two secondary maxima are clearly resolved experimentally for Ne-Na_2 at E = 175 meV (Fig. 4.1), although theory predicts eight for the same energy and j_2 = 6 [4.16]. In our opinion j_2-distributions are the more appropriate representation. This reflects the fact that j_2 is the conjugate variable to the orientation angle γ and, once again, it is the γ-dependence of the potential $V(R,\gamma)$ which causes the rotational rainbows.

4.3.5 Relationship to Other Work

The preceeding analysis of rotational rainbows is based on the classical limit of the quantum, energy sudden approximation in the so-called j_z-conserving approximation. Recently, the classical S-matrix theory of MILLER and MARCUS [4.23,24] has been expressed in the j_z-conserving approximation [4.34] without, however, making any further approximations. In that theory and using the "initial value representation" [4.34,36], the ℓ-dependent S-matrix for $j_1 \rightarrow j_2$ transition is given by

$$S^{\ell}(j_1 \rightarrow j_2) = (2\pi)^{-1} \int_0^{2\pi} d\bar{\gamma}_1 (\partial\bar{\gamma}_2/\partial\bar{\gamma}_1)^{\frac{1}{2}} \exp\{i[\phi(\bar{\gamma}_1) + \bar{\gamma}_2 j_2(\bar{\gamma}_1) - \bar{\gamma}_2 j_2]\} \quad . \tag{4.43}$$

The quantities $\bar{\gamma}_1$ and $\bar{\gamma}_2$ are basically the initial and final orientation angles of the rigid rotor, $j_2(\bar{\gamma}_1)$ is the final value of the rotational angular momentum as determined from a classical trajectory, and the phase $\phi(\bar{\gamma}_1)$ is given by the action integral

$$\phi(\bar{\gamma}_1) = - \int_{-\infty}^{\infty} dt \left[R(t) \frac{dP(t)}{dt} + \bar{\gamma}(t) \frac{dj(t)}{dt} \right] \tag{4.44}$$

where initially

$$\bar{\gamma}(t_1) = \bar{\gamma}_1 + 2B_{rot} j_1 \mu R(t_1)/P(t_1); \quad j(t_1) = j_1 \quad . \tag{4.45}$$

$P(t)$ is the momentum conjugate to the radial distance $R(t)$.

It is easy to show that in the sudden limit (4.43) reduces to the semiclassical IOS expression for $S^{\ell}(j_1 \rightarrow j_2)$. In that limit we have, by definition, the following results [4.37]:

$$\frac{d\bar{\gamma}(t)}{dt} = 0 \; ; \quad \frac{dj(t)}{dt} = 0 \tag{4.46a}$$

so that

$$\bar{\gamma}_1 = \bar{\gamma}_2 \quad ; \quad (\partial\bar{\gamma}_2/\partial\bar{\gamma}_1) = 1 \tag{4.46b}$$

$$j_2(\bar{\gamma}_1) = j_1 \tag{4.46c}$$

and the action reduces to

$$\phi(\bar{\gamma}_1) = - \int_{-\infty}^{\infty} dt \, R(t) \frac{dP(t)}{dt} = 2\eta(\bar{\gamma}_1) \tag{4.46d}$$

with $\eta(\bar{\gamma}_1)$ the JWKB phase shift (4.4) for fixed orientation $\bar{\gamma}_1$. Note that all trajectory quantities depend parametrically on the partial wave. Equation (4.46a) is the classical statement of the sudden limit and (4.46b-c) are all consequences of it. Inserting these results into (4.43) we obtain

$$S^{\ell}(j_1 \to j_2) = (2\pi)^{-1} \int_0^{2\pi} d\bar{\gamma}_1 \exp\{i[2\eta(\bar{\gamma}_1) + \bar{\gamma}_1(j_1 - j_2)]\} \quad (4.47)$$

which is identical to the sudden approximation result given previously by BOWMAN and LEE [Ref.4.26, Eq.(6b)]. The expression used in [4.25] differs from (4.47) because there (as in the present chapter) the classical limit for the initial $j_1 = 0$ state is *not* employed; instead the quantum expression is used. This was done because the classical limit for the ground rotational state is very inaccurate and leads to inaccurate cross sections [4.26]. Thus, the semiclassical IOS expression for $S^{\ell}(j_1 \to j_2)$ follows directly from the sudden limit of the initial value representation of the classical S-matrix. It should be noted that by treating the internal degrees of freedom in the sudden limit the need to explicitly integrate the classical equations of motion disappears. This of course accounts for the great simplicity and ease of interpretation of the semiclassical IOS approximation. We should also note that expressions similar to (4.47) appear in numerous semiclassical theories of inelastic scattering [4.38], all of which are approximations to classical S-matrix theory.

The subsequent complete (i.e., two-dimensional) stationary phase evaluation of the semiclassical sudden scattering amplitude had not been done previously in classical S-matrix theory and thus the rainbow analysis based on the vanishing of the determinant $D(\ell,\gamma)$, see (4.27), and all of the subsequent analysis is new. (An analogous rainbow analysis of collinear vibrationally inelastic scattering using classical S-matrix theory has been given, however [4.23,36].) Because that determinant is a classical quantity, a relationship does exist between it and some prior classical work done by THOMAS [4.39,40] and BECK et al. [4.41].

THOMAS [4.39] analyzed the results of his quasiclassical trajectory calculations of inelastic differential cross sections for Li^+ + CO and Li^+ + N_2 by considering the determinant of a Jacobian matrix which is the reciprocal of $D(\ell,\gamma)$ in (4.27). His important contribution was to note that rainbows could occur whenever his determinant became singular [4.39]. That condition is equivalent to our condition that $D(\ell,\gamma)$ vanishes. However, there are some differences in the possible interpretation of these conditions. As THOMAS [4.40] recently noted *each* element of his Jacobian matrix becomes simultaneously singular when the determinant is singular. Thus, it is not clear how, based on his analysis, to classify the rainbow into types. However, as shown previously [4.16,33] and above, a classification of rainbow types can be possible based on our analysis. A somewhat different classical analysis of rainbows has been given by BECK et al. [4.41] but for scattering by a hard shell potential. A discussion of their analysis is given in Sect.4.6.1.

Quite a different approach to discussing rotational rainbow scattering is presented in [4.42]. There the time-dependent Schrödinger equation for the rotational degree of freedom is solved along an average classical trajectory R(t) within a j_z-conserving approximation. The set of first-order differential equations coupled in j is solved exactly rather than by applying the energy sudden approximation which, if done, would yield S-matrix expressions similar to (4.15) or (4.47). The asset of this classical path, close coupling approach (see, for example, [Ref.4.43, Sect.3]) is its ease of interpretation because the time evolution of the rotational excitation along the trajectory can be examined. The advantage over quasiclassical trajectory calculations is that for a given impact parameter only a single "trajectory" has to be calculated and Monte Carlo averaging over initial angles is not necessary. Furthermore, all quantum effects are automatically incorporated.

4.4 Numerical Examples

4.4.1 Homonuclear Case

The He-CO system, with CO treated as a rigid-rotor, has been studied extensively by a variety of methods [4.34,44-49]. In several studies the interaction potential was chosen as

$$V(R,\gamma) = V_0(R)[1 + a_1P_1(\cos\gamma) + a_2P_2(\cos\gamma)] \tag{4.48}$$

where

$$V_0(R) = \varepsilon\left[\left(\frac{R_m}{R}\right)^{12} - 2\left(\frac{R_m}{R}\right)^6\right] \tag{4.49}$$

and $\varepsilon = 20.7\ \mathrm{cm}^{-1}$ and $R_m = 6.67$ bohr. In the classical S-matrix study (in the j_z-helicity conserving approximation) of MCCURDY and MILLER [4.34], the constant a_1 was varied from 0.0-0.8 and a_2 kept at 1.2. The case $a_1 = 0.0$ corresponds to a homonuclear diatom which we might label as N_2 instead of CO. It is this case we consider here. A quantal sudden (IOS) calculation of total and differential cross sections for rotationally inelastic scattering has been reported and an analysis of the rotationally rainbow scattering in the differential cross sections given [4.26,28].

Because this system is in several ways prototypical of group II systems where there is strong, short range, repulsive coupling between the rotational and translational degrees of freedom and a fairly weak long-range interaction, we choose it as a first numerical example of the general results described in the previous section.

First, let us demonstrate that the energy sudden approximation is indeed valid for this system. Consider an initial translational energy equal to 150ε and the rotational constant $B_{rot} = 0.1\varepsilon$. Thus, $E/B_{rot} = 1500$. A very stringent way of testing the validity of the sudden approximation is to compare the classical limit sudden approximation deflection functions $\chi(\ell,\gamma)$ and $J(\ell,\gamma)$, given by (4.25,26) in the previous section with those computed from classical trajectories. This test has already been reported for the $J(\ell = 10,\gamma)$ excitation function for $a_1 = 0.0$

and $a_2 = 1.2$ [4.50]. The sudden approximation $J(\ell = 10,\gamma)$ was shown to be in excellent agreement with the classical trajectory result of MCCURDY and MILLER [4.34]. The agreement between the sudden approximation and classical trajectory $\chi(\ell = 10,\gamma)$ deflection function is also excellent [4.50]. At the integral cross section level the semiclassical IOS results using (4.47) were in excellent agreement with the initial value representation classical S-matrix cross sections [4.34]. Thus, we conclude that the He-(homonuclear)CO scattering is very well described by the sudden approximation (for $E/B_{rot} = 1500$, at least).

Deflection Functions and Rainbow Curves

The sudden approximation excitation function $J(\ell,\gamma)$ is plotted versus γ for five values of ℓ in Fig.4.6a. As anticipated in Sect.4.3.4, the amplitude of the $J(\ell,\gamma)$, shown in Fig.4.6b decreases as ℓ increases. Note that except for large ℓ the maximum in each $J(\ell,\gamma)$ curve occurs at nearly the same value of γ (65°, approximately). This particular value was labeled γ_R in the previous section because here the stationary phase roots to (4.26) coalesce for some (continuous) value of j_2 and to a good approximation the classical cross section diverges at a particular scattering angle $\theta \simeq \chi(\ell,\gamma_R)$. Below a comparison of the rainbow curve, $j_R(\theta)$ is made against $j_{max}(\theta)$ from the quantal IOS calculations and $j_R(\theta)$ obtained with the approximate determinant (4.28) instead of the exact one in (4.27). Before doing that, consider Fig.4.7 where the deflection function $\chi(\ell,\gamma)$ is plotted against ℓ for $\gamma = 0^\circ$ and $\gamma = 90^\circ$. As seen in Fig.4.7a for a large range of ℓ, $\ell \lesssim 60$, the variation of χ with γ is small compared to its variation with ℓ. Also, the $\chi(\ell,\gamma)$ all display a shallow minimum at $\ell \sim 95$ (see Fig.4.7b). At this value of ℓ, however, all inelastic transitions are highly classically forbidden (as can be surmised by inspection of Fig.4.6b). This minimum would be expected to produce a standard rainbow feature in the elastic differential cross section; however, it would occur at very small scattering angles and probably be unobservable. Inspecting Fig.4.7b we note that the condition $(\partial\chi/\partial\gamma)_\ell \ll (\partial\chi/\partial\ell)_\gamma$, which allows us to distinguish between different rainbow types, is obviously *not* fulfilled in the minimum region.

A comparison of the rainbow curves $\theta_R(j_2)$, i.e., the inverse of $j_R(\theta)$, obtained from the vanishing of the full determinant (4.27) and the approximate one (4.28) along with the $\theta_{max}(j_2)$ curve from the quantal IOS differential cross sections is given in Fig.4.8. The horizontal "error bars" at the $\theta_{max}(j_2)$ points serve to indicate the breadth of the main quantum rainbow maxima. Within this breadth the quantum differential cross section is larger than 90% of its value at the rotational rainbow maximum. Note that the maxima broaden as j_2 increases. The two classical rainbow curves $\theta_R(j_2)$ are in excellent agreement with each other, confirming the accuracy of the approximation in (4.28). Also, note that the shapes of the $\theta_R(j_2)$ and $\theta_{max}(j_2)$ curves as well as their divergence from each other are very similar to the curves shown already in Fig.4.5. According to (4.41,42), the difference

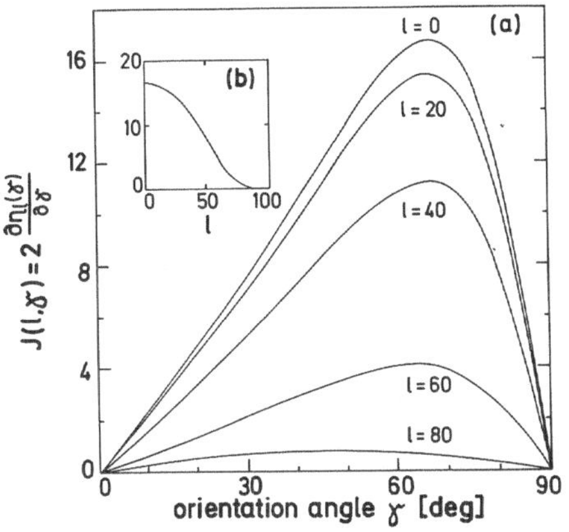

Fig.4.6 ▲

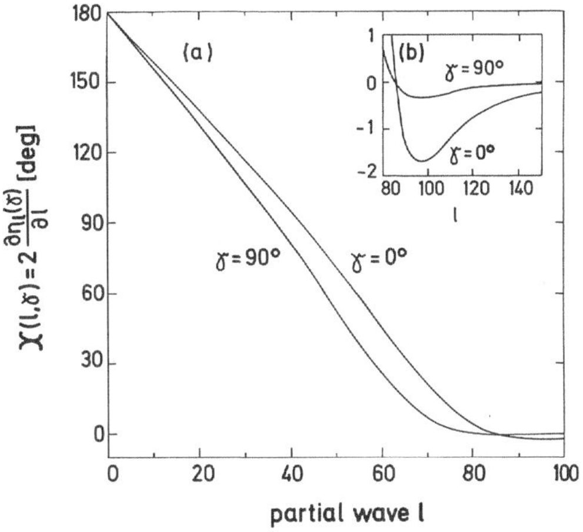

Fig.4.7 ▶

Fig.4.6. (a) Excitation function $J(\ell,\gamma)$, (4.26), versus γ for selected partial waves ℓ for the model He-N_2 system. The potential energy surface is defined in (4.48,49) and the collision energy is $E = 150\varepsilon$. (b) The maximum of $J(\ell,\gamma)$ versus γ as a function of ℓ

Fig.4.7. (a) Deflection function $\chi(\ell,\gamma)$, (4.25), versus ℓ for $\gamma = 0^\circ$ and $\gamma = 90^\circ$ for the model He-N_2 system. (b) Enlargement of the minimum region at large values of ℓ

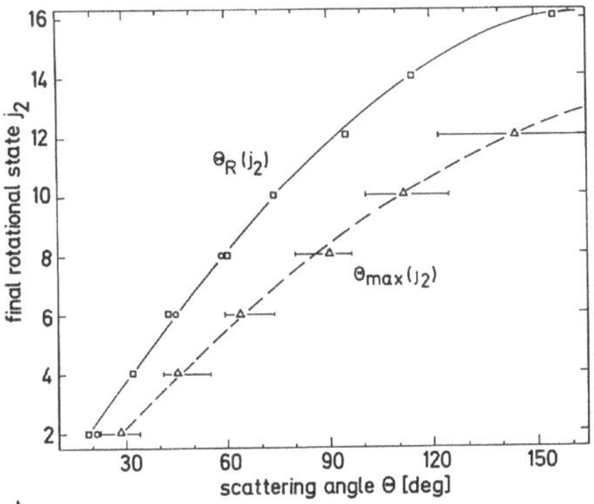

▲

Fig.4.8. Classical (□) and quantal (Δ) rainbow curves $\theta_R(j_2)$ and $\theta_{max}(j_2)$, respectively, for the model He-N_2 system at $E = 150\varepsilon$. Also shown are the classical results (o) obtained using the approximate determinant (4.28) instead of the exact one (4.27). The "error bars" on the quantal data points indicate the "breath" of the rainbow maxima

between $j_{max}(\theta)$ and $j_R(\theta)$ increases monotonically with the absolute value of $\partial^2 J/\partial\gamma^2$ evaluated at $\gamma = \gamma_R$. Inspection of Fig.4.6a shows that this second derivative increases with j_2. Thus, the difference between $j_R(\theta)$ and $j_{max}(\theta)$ grows with increasing j_2 which in turn increases with θ (Fig.4.8). Thus, we conclude that the difference between the classical singularity and the quantal rainbow maximum increases with the scattering angle.

Cross Sections and Probabilities

The quantum IOS differential cross sections $d\sigma/d\Omega(0 \rightarrow j_2|\theta)$ are plotted against the scattering angle θ for $0 \leq j_2 \leq 16$ in Fig.4.9a and against j_2 for $\theta = 60^\circ$, 120°, and 180° in Fig.4.9b. Consider Fig.4.9a first. The main maxima in the dif-

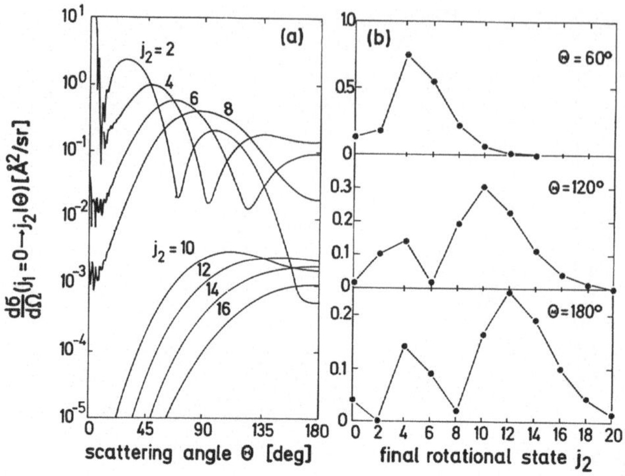

Fig.4.9a,b. IOS differential cross sections for the model He-N_2 system at E = 150ε. (a) Angular distributions for fixed j_2 the $j_2 \geq 10$ cross sections are multiplied by 10^{-2}; and (b) j_2-distributions for fixed scattering angle θ

ferential cross sections constitute the rainbow curve $\theta_{max}(j_2)$ plotted in Fig. 4.8. As already noted, these maxima occur at larger scattering angles and broaden as j_2 increases. For $j_2 = 2$ there is, in addition to the primary rotational rainbow maximum, a secondary supernumerary maximum at $\theta \simeq 100^{\circ}$. Secondary rainbow maxima are only incompletely developed for $j_2 = 4$ and 6. They are not observed for $j_2 \geq 8$. The j_2-distributions at $\theta = 120^{\circ}$ and 180° also display a primary and secondary rainbow maximum. The primary maximum shifts to lower j_2 values with decreasing θ, in qualitative accord with the behavior of the quantal and classical rainbow curves shown in Fig.4.8. It should be stressed again that the two representations of differential cross sections in Figs.4.9a,b are equivalent [4.21].

It was stated in the previous section that to a very good approximation and especially in the backward direction, each scattering angle θ can be assigned to one partial wave ℓ for the class of collision systems we are considering here. That would imply that Fig.4.9b can be reproduced by replaxing θ by ℓ(θ) and the differential cross sections by the ℓ-dependent transition probabilities [4.21,29]. The ℓ-values corresponding approximately to the θ-values of Fig.4.9b are determined from Fig.4.7. The quantal IOS partial wave transition probabilities are given by

$$P^{\ell}(0 \to j_2) = |T^{\ell}_{m=0}(0 \to j_2)|^2 \qquad (4.50)$$

with the IOS T-matrix defined in (4.15). They are plotted in Fig.4.10b against j_2 for the values $\ell = \ell(\theta)$. Comparing this figure with Fig.4.9b, a striking similarity is seen, as predicted. The $P^{\ell}(0 \to j_2)$ are plotted against ℓ in Fig.4.10a. The maxima in the curves for $j_2 > 2$ correspond to the primary rainbow in the respective differential cross section. The $P^{\ell}(0 \to 2)$ has three maxima; the main one at

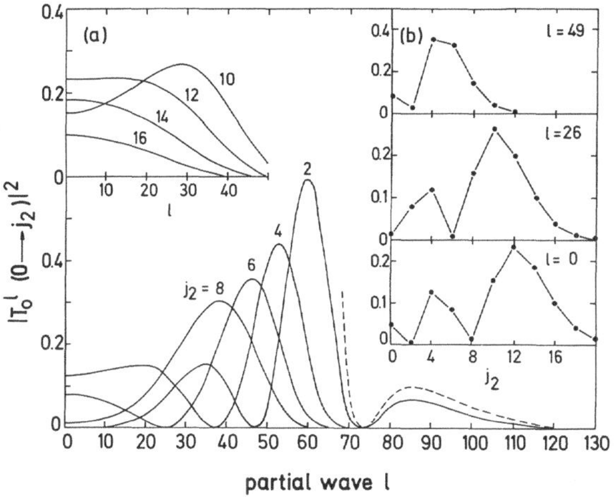

Fig.4.10. (a) IOS $0 \to j_2$ partial transition probabilities,(4.50), versus partial wave ℓ for the model He-N_2 system at $E = 150\varepsilon$. The dashed curve is the elastic $j_1 = 0 \to 0$ result. (b) The same results plotted versus j_2 for fixed partial waves $\ell = 49$, 26 and 0 corresponding roughly to the angles of Fig.4.9b, i.e., $\theta = 60°$, 120° and 180°

$\ell \simeq 60$ corresponds to the primary rotational rainbow maximum in the differential cross section at $\theta \simeq 28$ and the smaller one at $\ell \simeq 35$ corresponds to a secondary, supernumerary rotational rainbow maximum at $\theta \simeq 100°$. The third and smallest maximum at $\ell \simeq 85$ is not associated with a rotational rainbow. It corresponds instead to the maximum in the quantity $(2\ell + 1)P^{\ell}(0 \to 0)$ (which classically is predicted to occur when $d\chi/d\ell$ vanishes). In the present case this occurs for $\ell \simeq 95$. The $P^{\ell}(0 \to 0)$ maximizes at a values of ℓ less than 95, of course; in the present case at $\ell \simeq 85$, as shown by the dashed curve in Fig.4.10a. This additional feature in the $P^{\ell}(0 \to 2)$ is quite intriguing because the $0 \to 2$ transition is classically forbidden at these partial waves, yet $P^{\ell}(0 \to 2)$ exhibits an oscillation. The $P^{\ell}(0 \to 4)$ also has a very small local maximum at $\ell \simeq 85$, however, too small to be seen in the figure. Evidently, these probabilities tunnel into that region in ℓ-space where the $d\chi/d\ell = 0$ rainbow causes them to "jump-up" slightly before decaying to zero.

Comparison Between Quantal and Semiclassical IOS Cross Sections

As stated in Sect.4.3.3, we will demonstrate that the uniform semiclassical approximation to the IOS scattering amplitude is not only qualitatively but also quantitatively correct. In Table 4.1 we list $0 \to j_2$ differential cross sections for He-(homonuclear)CO scattering as obtained from the classical (4.40), the primitive semiclassical, (4.39), the uniform semiclassical (4.37,38) and the exact quantal IOS approximation. The potential parameters are the same as defined after (4.49) and the collision energy is $E = 150\varepsilon = 0.385$ eV. A zero classical cross section indicates that the transition is classically forbidden at that particular scattering angle.

We note the following observations: (a) the uniform semiclassical IOS cross sections agree very well with the quantal ones. The only exception is the elastic

Table 4.1. Comparison between classical, semiclassical and quantal IOS differential cross sections (in Å^2/sr) for $0 \to j_2$ transitions in He-(homonuclear)CO collisions. The collision energy is $E = 150\varepsilon = 0.385$ eV

j_2	classical; (4.40)	primitive semi-classical; (4.39)	uniform semi-classical; (4.37,38)	quantal
$\theta = 50^\circ$				
0	0.250	0.010	0.007	0.015
2	0.538	0.792	0.812	0.770
4	0.637	1.103	0.932	0.947
6	1.246	1.315	0.418	0.421
8	0	0.159	0.125	0.126
10	0	0.332	0.030	0.030
12	0	0.007	0.006	0.006
$\theta = 100^\circ$				
0	0.053	0.014	0.015	0.009
2	0.108	0.209	0.207	0.205
4	0.149	0.034	0.030	0.032
6	0.175	0.147	0.158	0.154
8	0.199	0.374	0.361	0.360
10	0.243	0.388	0.301	0.302
12	0.567	0.586	0.154	0.155
14	0	0.078	0.059	0.059
16	0	0.021	0.018	0.019
18	0	0.005	0.005	0.005
$\theta = 150^\circ$				
0	0.033	0.054	0.053	0.046
2	0.062	0.007	0.008	0.007
4	0.086	0.164	0.164	0.163
6	0.104	0.064	0.059	0.061
8	0.118	0.040	0.045	0.044
10	0.132	0.209	0.212	0.210
12	0.153	0.286	0.257	0.257
14	0.206	0.277	0.175	0.176
16	0	0.195	0.085	0.086
18	0	0.041	0.033	0.033
20	0	0.012	0.011	0.011
22	0	0.003	0.003	0.003

$j_1 = 0 \to 0$ transition which, however, could be forseen because the classical limit (large index) of the rotor wave function is certainly not correct for the ground state. (b) The primitive semiclassical IOS cross section (4.39) agrees well with the uniform cross section for transitions not too close to the rainbow state, $j_R(\theta)$. Using the asymptotic (large argument) limit of the Airy-functions, MILLER [4.36] showed that the primitive semiclassical expression follows directly from the Airy-uniform expression for large differences of the phases associated with the interfering trajectories. (c) Both the classical and primitive semiclassical cross sections diverge at the rainbow state. The classical cross section rises monotonically and averages over the quantal interference oscillations. The numerical example for a heteronuclear system in [4.27] proves that the quantitative accuracy of the uniform IOS approximation is not restricted to homonuclear molecules only.

Energy Dependence

So far we have discussed rotational rainbow scattering for a fixed energy only. Next, we investigate the dependence of rotational rainbow features on the collision energy E. For this purpose we choose another homonuclear system, He-Na_2, for which an accurate *ab initio* potential surface has been calculated [4.51]. With a rotational constant of 1.9183×10^{-2} meV and an (average) well depth of the van der Waals minimum of about 0.1 meV, this system is also prototypical for studying rotational rainbow scattering (for a comparison between theoretical and experimental differential cross sections see the next section). In Fig.4.11 we plot quantal IOS differential cross sections versus j_2 for $\theta = 150^\circ$ and three energies (the corresponding classical rainbow curves $j_R(\theta)$ are shown in Fig.5 of [4.27]). With increasing energy the main rotational rainbow maximum j_{max} is shifted to larger j_2. Simultaneously, successively more supernumerary rainbows arise in the classically allowed region.

The positions of the main rainbow maxima j_{max} are plotted versus $E^{\frac{1}{2}}$ in the inset of Fig.4.11. Except for the highest energies we observe an almost linear dependence, i.e., $j_{max} \sim E^{\frac{1}{2}}$. The same behavior has been found in an IOS study for model K-N_2 and CO systems [4.29]. The simple model of scattering of an atom from a hard shell molecule (see also Sect.4.6.1) predicts for the classical rainbow state [4.52,53]:

$$j_R(\theta,E) = 2ck\,\sin\theta/2 = 2c(2\mu E)^{\frac{1}{2}}\sin\theta/2 \qquad (4.51)$$

where the constant c depends exclusively on the potential shell. Note, this ultra simple formula explains qualitatively the angular dependence of j_R as anticipated in Fig.4.5a and observed for a realistic system in Fig.4.8. It also describes almost exactly the energy dependence of j_{max} at fixed scattering angle shown in Fig. 4.11. A linear dependence as indicated by (4.51) has also been derived for a model but realistic potential in [4.29] considering backward scattering $\theta = 180^\circ(\ell = 0)$. The scaling of j_{max} with mass, energy, and scattering angle as predicted by (4.51) is recently confirmed experimentally by JONES et al. [4.13b] comparing the results for Ne, Ar, Kr, and Xe colliding with Na_2 including transitions up to $\Delta j = 16$.

Slight deviations from the simple $E^{\frac{1}{2}}$ behavior of $j_{max}(\theta)$ are indicated in Fig. 4.11 for the highest energies. They have also been observed experimentally for the K-N_2 and CO systems [4.9] and found in a recent theoretical study for Ne-Na_2 [4.16]. In the latter study we tried to relate these findings to special features of the potential surface. We argued that a linear E dependence is surmised if the potential can approximately be written in the factorized form

$$V(R,\gamma) = h(R)g(\gamma) \qquad (4.52)$$

which has been utilized in the model study of SCHINKE [4.29]. Inspection of Fig. 2 in [4.51] clearly shows that (4.52) is not appropriate to represent the He-Na_2 surface above $V \simeq 0.2$ eV (see SCHINKE et al. [4.16] for a more detailed discussion).

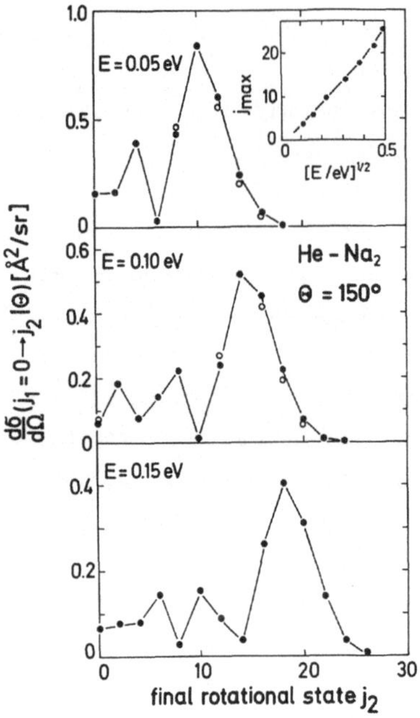

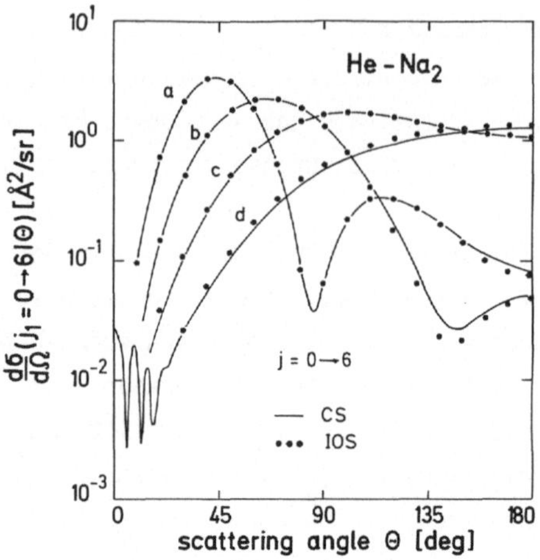

Fig.4.12. IOS (●●●) and CS (——) differential cross sections versus θ for $j_1=0 \to 6$. The collision system is He-Na_2 [4.51]. (a) E=100 meV, (b) E=50 meV, (c) E=25 meV and (d) E=12.5 meV

Fig.4.11. IOS (●) and CS (o) differential cross sections versus j_2 for $\theta = 150^\circ$ and three collision energies. The collision system is He-Na_2 [4.51]. The inset shows the quantal rainbow state $j_{max}(\theta=150^\circ)$ versus $E^{1/2}$

The shift of $j_R(\theta)$ and likewise $j_{max}(\theta)$ with E at each scattering angle implies that for each transition the rainbow maximum occurs at successively lower scattering angles if E is increased, i.e., $\theta_R(j_2)$ and $\theta_{max}(j_2)$ are decreasing functions of E. This is illustrated in Fig.4.12 showing the IOS differential cross section for $j_2 = 6$ at four different energies. The diffraction oscillations due to scattering from the highly repulsive core of the potential energy surface are very narrow and restricted to small scattering angles. They are only shown for the lowest energy to make the presentation clearer. For E = 12.5 meV, the $0 \to 6$ transition is classically forbidden over the entire range of scattering angles and the cross section rises monotonically out of the forward direction. A very broad rainbow maximum is built up at $\theta_{max} \simeq 100^\circ$ for E = 25 meV. With increasing energy the rainbow maximum shifts into the forward direction, becomes narrower and its magnitude increases. Simultaneously, a supernumerary maximum arises in the classically allowed angular region and it is fully developed for the highest energy, E = 100 meV, at $\theta \simeq 115^\circ$. Monotonically increasing differential cross sections similar to curve (d) in Fig.4.12 have been determined experimentally for $0 \to 1$ and $0 \to 2$ transitions in Ne-HD [4.3] and Ne-D_2 [4.4] collisions which belong to group I of Sect.4.1.1. Investigating the variation of the differential cross section with the collision energy, we surmise that rotational rainbow structures may be characteristic for *any* system, even those of group I, provided the energy is high enough to open many states and the rotational coupling is sufficiently strong to cause their excitation.

Also shown in Figs.4.11,12 are differential cross sections calculated in the coupled-states (CS) approximation, which fulfils energy conservation in contrast to the IOS approach. In each case we observe almost exact agreement, which certainly proves that the energy sudden condition is valid for the transitions and energies considered. A detailed comparison between IOS and CS cross sections for Ne-Na_2 is given in [4.16]. The centrifugal sudden condition inherent to both approximations is expected to be valid because all collisions except those leading to very small scattering angles (a region of no interest for our present purposes) are dominantly repulsive. The ratio of the collision energy to the well depth is at least 100. Therefore, the IOS cross sections for He-Na_2 are regarded as almost exact. This conclusion is important for the comparison between theoretical and experimental cross sections which we shall discuss in Sect.4.5.2.

4.4.2 Heteronuclear Case

Transitions Out of the Ground State

According to the general discussion in Sect.4.3.4, especially Fig.4.3, we expect even more abundant cross section features for scattering with heteronuclear molecules. In particular we predicted bi-modal j_2-distributions, which have indeed been observed in two experiments [4.9,10,17], and more complicated oscillation patterns. In this section we present some differential cross section calculations for heteronuclear target molecules using a model potential surface [4.29] which crudely represents the K-CO scattering system. Figure 4.13 shows differential cross sections versus final state j_2 at three selected scattering angles. The scattering energy is E = 1.24 eV. As predicted in Sect.4.3.4, each distribution shows two distinct rainbow maxima at $j_{max}^{(1)}$ and $j_{max}^{(2)}$, respectively. With decreasing scattering angle both maxima shift to lower transitions and move together. The strong oscillations below $j_{max}^{(1)}$ are due to the interference of four stationary phase points in (4.29b). Each pair of them represents scattering from the C or the O-end. In all cases the first rainbow maximum is modulated by rapid oscillations which slowly die off with increasing $j_2 > j_{max}^{(1)}$ because the contribution from the second branch in Fig.4.3b [A_{34} in (4.29)] becomes negligibly small compared to A_{21}. A supernumerary maximum is clearly observed at $j_2 = 44$ for $\theta = 150^{\circ}$. It is due to constructive interference between γ_1 and γ_2 in Fig.4.3b which belong to the same branch of the excitation function $J(\ell,\gamma)$. The second rainbow maximum is finally smooth, although small oscillations are indicated at the lowest angle when both rainbow maxima are close together and interference between (γ_1,γ_2) and (γ_3,γ_4) is still important.

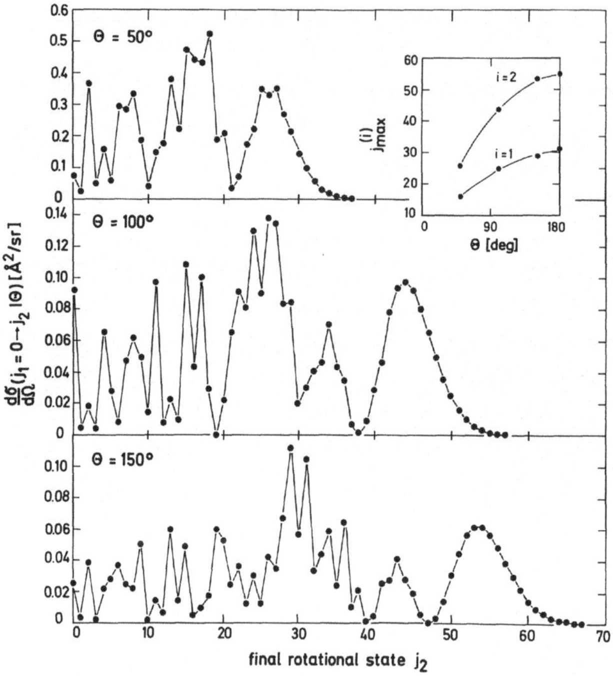

Fig.4.13. IOS differential cross sections versus j_2 at fixed scattering angle θ for a model K-CO system [4.29]. The collision energy is E = 1.24 eV. The inset shows the quantal rainbow curves $j_{max}^{(i)}(\theta)$, i = 1, 2

Within the semiclassical IOS approximation of scattering from hard shell potentials [4.53] it is readily shown that (see also Sect.4.6.1)

$$j_R^{(i)}(\theta) = 2c^{(i)}k \sin\theta/2 \quad , \quad i = 1, 2 \tag{4.53}$$

is the extension of (4.51) for heteronuclear molecules. Equation (4.53) predicts that each rainbow branch $j_R^{(i)}$ is proportional to $\sin\theta/2$. Such behavior is qualitatively shown for the quantum mechanical rainbow maxima $j_{max}^{(i)}$ in the inset of Fig. 4.13. Furthermore, at each scattering angle both $j_R^{(1)}$ and $j_R^{(2)}$ are proportional to $E^{\frac{1}{2}}$. Thus, increasing the collision energy enlarges the shift between both rainbow maxima. This was found empirically [4.29] and observed experimentally by SCHEPPER et al. [4.9]. There is yet another interesting implication of (4.53). The isotopic substitution of one atom of the target molecule changes the reduced mass μ *and* the asymmetry of the potential surface, i.e., the constants $c^{(i)}$ in (4.53). The net effect for fixed θ and E is a shift of the two rainbow maxima. This has been observed by BECK et al. [4.10] in nice experiments with K - $C^{18}O$ and K - $C^{16}O$, respectively. Qualitatively the same shift is predicted by IOS calculations [4.29].

Transitions Out of Excited States

So far we have considered only scattering out of the rotational ground state $j_1 = 0$, a situation which is normally not met in experiments unless the target beam is rotationally very cold. The influence of excited initial states on rotational rainbow structures is investigated in Fig.4.14 showing differential cross sections at $\theta = 150^\circ$ for $j_1 = 1$, 3 and 5 using again the model potential surface of [4.29]. These cross sections are easily obtained from the $j_1 = 0$ ones utilizing the factorization formula (4.18). First, we note that the rainbow maxima $j^{(1)}_{max}$ and $j^{(2)}_{max}$ are approximately shifted from the $j_1 = 0$ ones (shown in Fig.4.13) by the amount $j_1 = 1$, 3 or 5. This indicates that $\Delta j = |j_2 - j_1|$ rather than the final state j_2 alone is the proper conjugate variable to the orientation angle γ [4.28]. The same conclusion is derived from an extensive theoretical investigation for Ne-Na_2 [4.16] and confirmed experimentally for the same system [4.14] (Fig.4.19).

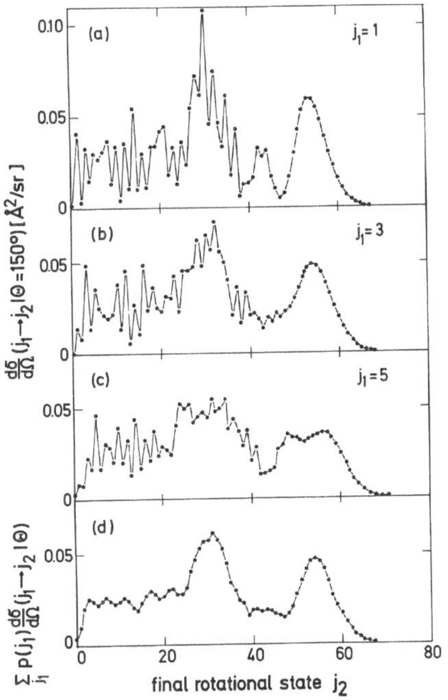

Fig.4.14. IOS differential cross sections versus j_2 at fixed scattering angle $\theta = 150^\circ$ for a model K-CO system [4.29]. The collision energy is E = 1.24 eV. (a) $j_1 = 1$, (b) $j_1 = 3$ and (c) $j_1 = 5$. The distribution in (d) represents a weighted average over j_1 which roughly represents the experimental condition [4.9]

Secondly, one observes that the oscillation structure is gradually damped as the initial state is increased. The reason is simply that $(2j_1 + 1)(2j_2 + 1)$ cross sections with different initial and final magnetic states m_1 and m_2 are summed in (4.17) to yield the degeneracy-averaged cross section $d\sigma(j_1 \to j_2|\theta)/d\Omega$. Even with the simple IOS scattering amplitude in (4.14), which is diagonal in m, $(2j_1 + 1)$ cross sections with quite different oscillatory behavior contribute to $d\sigma(j_1 \to j_2|\theta)/d\Omega$. In many experiments, measured differential cross sections represent an average over a broad distribution of initial rotational states j_1. Figure 4.14d shows the result of such an average using IOS cross sections with $0 \le j_1 \le 6$. The two main rainbow maxima are still clearly observable but the interference structures are almost completely washed out. This damping is likewise observed for homonuclear molecules [4.29] and angular distributions of single Δj transitions ([4.16] and Fig.4.19). Thus, we conclude that only state-to-state experiments with low initial rotational states, ideally the ground state, have a realistic chance of resolving both the main rotational rainbow maxima and the interference structures, the supernumerary rainbows, for example. Indeed, supernumeraries are still only resolved for $0 \to 2$, 4 and 6 transitions in Ne-Na_2 collisions (see Fig.4.1). These data are therefore the first real proof of the existence of secondary rainbows because exact close coupling calculations are probably unmanageable under strong coupling conditions.

4.5 Recent Experiments and Comparison with Theory

Basically two quite different methods are, in general, utilized to resolve rotational transitions in inelastic scattering experiments [4.1,2]: (i) the *velocity change* method (for example, time-of-flight measurements) and (ii) *state selective* techniques (for example laser induced fluorescence). The first method takes advantage of the velocity change due to the reduced energy transfer

$$\Delta E/E = B_{rot}[j_2(j_2 + 1) - j_1(j_1 + 1)]/E \tag{4.54}$$

accompanying a $j_1 \to j_2$ transition. It works best if B_{rot} is of the same order as the initial translational energy E and it has successfully been applied to group I systems. (Note that the recent measurements of FAUBEL et al. [4.8] showed that method (i) is applicable for ratios E/B_{rot} as large as 100.) However, as discussed in the Sect.4.1.1, rotational rainbow scattering is most pronounced for large ratios E/B_{rot} when many states are energetically accessible, which makes the resolution of single transitions by methods based on the velocity change extremely difficult, if possible at all. Thus, only the gross features, i.e., the main rotational rainbow maxima, are expected to be observed while finer structures such as supernumerary maxima will very likely be obscured by the insufficient state resolution. In addition, the velocity change method is inherently unable to distinguish between energy transfer caused by different inelastic processes such as rotational or vibrational excitation and so a unique interpretation of the original spectra might be questionable.

All these disadvantages do not apply to the second method because it is not based on (4.54). Combined with a special technique to modulate the target beam [4.11-15], it guarantees state-to-state measurements. Thus, state selective methods are most promising for exploring *all* rotational rainbow features, provided the angular resolution is sufficient and the spread of the initial translational energy is small.

4.5.1 State-Unresolved Experiments

Li^+-N_2 and Li^+-CO

One of the earliest experiments indicating the existence of rotational rainbow maxima in final rotational state distributions at fixed scattering angles is that of BÖTTNER et al. [4.54]. These authors reported gaussian-shaped rotational distributions for scattering angles around 40° obtained from time-of-flight spectra for Li^+-N_2 and CO at E = 4.23 eV and 7.07 eV. The peaks at about $20 \lesssim j_2 \lesssim 30$, depending on the energy and the angle, are probably caused by constructive interference as described above. However, the overall shape of the distributions [Ref. 4.54, Figs.5 and 6]; the decay from the maximum to higher states is much slower than expected for classically forbidden processes) and the observation of a single maximum in the case of CO differs from the general predictions of Sect.4.3 and from other experiments and calculations as well.

Stimulated by the availability of these experimental data, *ab initio* potentials have been calculated for Li^+-N_2 and CO [4.55,56] to perform scattering calculations [4.39,56-60]. However, neither of these calculations has so far been able to explain the experimental results, consistently yielding distributions that are too narrow and peak at smaller transitions than observed experimentally. The situation for the Li^+-CO system has recently been reviewed by THOMAS [4.61]. Analyzing quasi-classical trajectory calculations for Li^+-CO, he was the first to speculate that rainbow-like structures might be observable in final state distributions [4.39]. Guided by this prediction, EASTES et al. [4.62] repeated the Li^+-CO experiment at smaller scattering angles $3.75^\circ \lesssim \theta \lesssim 12.5^\circ$. Indeed, they found evidence for structures in the time-of-flight spectra, which are clearly visible in the final state distributions. However, a recent classical study based on 156,000 trajectories was not able to describe either the j_2-distribution or the simulated time-of-flight spectrum constructed from the trajectory data and account for the finite resolution of the experimental apparatus [4.63]. Since the potential energy surface is believed to be highly accurate, THOMAS et al. [4.63] argued that quantum effects are responsible for this severe disagreement. This would not be inconsistent with the theoretical results of the previous section which showed that the classical (IOS) rainbow analysis may not be in quantitative agreement with the quantum or uniform analysis.

In our opinion, an analysis of rotational rainbows as presented in this chapter is not possible for the ionic systems Li^+-N_2 and CO. The attractive, long-range forces and the considerably deep potential well which, in addition, is strongly anisotropic, hinders a systematic classification into different rainbow types as suggested in Sect.4.3.2. Furthermore, a thorough theoretical study is very difficult because close-coupling calculations are not feasible at all and decoupling methods like the IOS and CS approximations may be too inaccurate. The situation would be clearer for larger scattering angles which dominantly sample the repulsive part of the potential. However, this region has not been explored either experimentally or theoretically.

K-N_2 and K-CO

SCHEPPER et al. [4.9] were the first to unambiguously resolve rotational rainbow maxima in Δj-distributions obtained from reduced recoil velocity distributions of potassium atoms scattered inelastically from N_2 and CO molecules. The approximate well depths of 7 meV for N_2 and 9 meV for CO [4.9] are negligibly small compared to the collision energies $0.39 \text{ eV} \leq E \leq 1.24 \text{ eV}$. This and wide angle scattering ($\theta \geq 90^{\circ}$) assure us that the collisions are dominantly repulsive. High resolution is obtained by computer controlled variation of laboratory reduced velocity and scattering angle such that the intensity of the scattered particles are directly measured in the centre-of-mass system.

In Fig.4.15 we show two original distributions versus reduced recoil velocity

$$u^*(j_2) = [1 - \Delta E(j_2)/E]^{\frac{1}{2}} \tag{4.55}$$

with $\Delta E/E$ defined in (4.54). The corresponding final rotational states

$$j_2 \simeq [E/B_{rot}(1 - u^{*2})]^{\frac{1}{2}} \tag{4.56}$$

valid for $j_2 \gg 1$ are given on the upper axis in each figure. In both cases one notices a sharp peak for almost elastic collisions, $u^* \simeq 1$. This maximum is simply due to the increased density of states in this energy range and is *not a dynamical feature*. It vanishes if the data are presented versus final rotational state j_2 rather than u^* (an example is given in [Ref.4.9, Fig.7]). The K-N_2 spectrum shows a distinct peak, the rotational rainbow maximum at $u^* \simeq 0.8$ ($j_2 \simeq 42$) beyond which the intensity rapidly falls off into the classically forbidden region to lower values of u^*. Note, at the energy of 1.24 eV, rotational transitions up to $j_2 \simeq 70$ are energetically open. For CO two rotational rainbow maxima are observed, the first one at $u^* \simeq 0.9$ ($j_2 \simeq 34$) and the second at $u^* \simeq 0.65$ ($j_2 \simeq 55$). The largest energetically accessible state for this system is $j_2 \simeq 65$.

Since accurate potentials are not available for K-N_2 and K-CO, a quantiative comparison between experiment and theory has not yet been done. IOS calculations have been performed using simple model surfaces [4.29] and the potential parameters are chosen to obtain qualitative agreement with the experimental results. The energy sudden condition is probably not valid for these systems where $\Delta E/E$ exceeds 60%, and thus these early IOS calculations should not be considered quantitative. Nevertheless, they explained for the first time rotational rainbow structures as observed experimentally in terms of approximate quantum scattering theory. The depen-

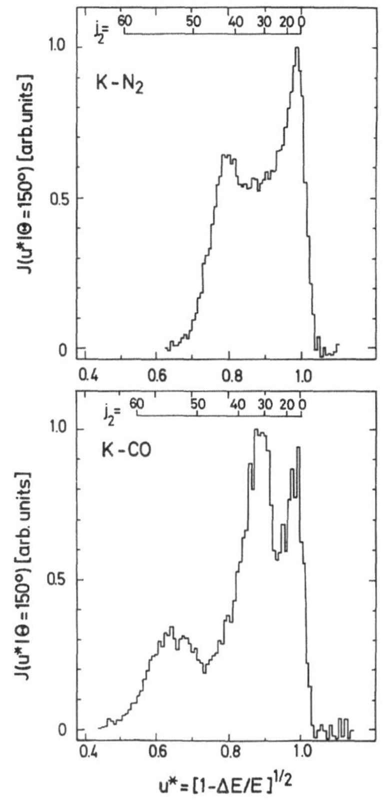

Fig.4.15. Experimental center-of-mass cross sections as a function of reduced recoil velocity u^* (4.55) of K atoms scattered at $\theta=150^\circ$ and $E=1.24$ eV from N_2 and CO. The corresponding final states j_2 are given on the upper scale in each figure (adapted from SCHEPPER et al. [4.9])

dence of the rotational rainbow maxima on scattering angle and collision energy were described in fair agreement with the experimental observations.

Interference structures in the classically allowed Δj region as predicted in Sect.4.3 and shown in Fig.4.13 are not resolved in these experiments, mainly because the state resolving power is not sufficient. Moreover, the cross sections in Fig.4.15 are the average over a relatively broad initial distribution with $j_1 \simeq 2.5$ as the mean value [4.9]. As demonstrated in Fig.4.14d, even the theoretical distribution averaged over j_1 as described in the caption to this figure shows only the main rainbow maxima whereas the finer structures are almost completely smeared out. Bi-modal energy loss distributions were recently reported by ANDRES et al. for D_2-CO at E = 87.2 meV [4.17].

4.5.2 State-Resolved Experiments

Up to now fully state-resolved differential cross sections for systems belonging to group II of the Sect.4.1.1 have been reported for the systems X-Na_2 only: X = He [4.11,12], X = Ne [4.13,14] and X = Ar [4.13,15]. Theoretical cross sections have been compared to experimental data for He and Ne-Na_2, for which accurate *ab initio* potentials have been determined [4.16,51]. These surfaces are highly aniso-

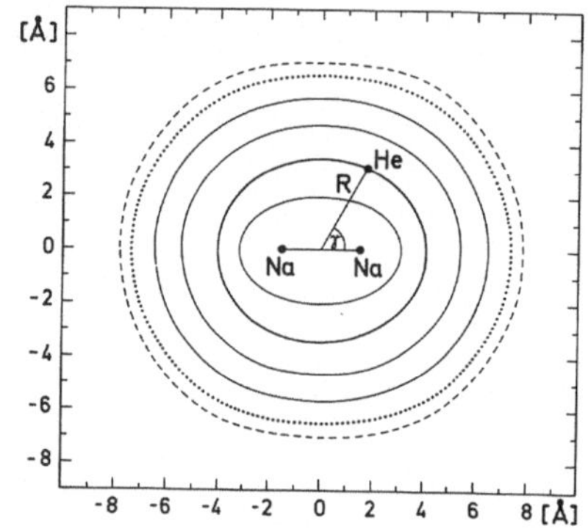

Fig.4.16. He-Na_2 interaction potential $V(R,\gamma)$ in polar coordinates. The plotted lines correspond to $V(R,\gamma) = 0.9 \times 10^i$ eV (i = 0, -1, -2, -3), 0.0 eV (dotted line) and -0.00009 eV (broken line) (adapted from SCHINKE et al. [4.51])

tropic and strongly repulsive. The calculated (average) well depth of the van der Waals minimum is $\varepsilon \simeq 0.1$ meV for He and $\varepsilon \simeq 0.3$ meV for Ne. They are negligibly small compared to the experimental collision energies of E = 90 meV and 175 meV, respectively. This and the extremely small rotational constant of 1.9183×10^{-2} meV makes these systems prototypical for studying rotational rainbow scattering. A contour plot of the He-Na_2 surface in polar coordinates is shown in Fig.4.16. To a good approximation each contour is well described by an ellipse. This fact may explain the success of the ultra simple hard ellipsoid model in describing, at least qualitatively, rotational rainbow scattering (Sect.4.6.1).

He-Na_2

We start the comparison between theory and experiment with the He-Na_2 system. The IOS differential cross sections have been transformed into the laboratory (LAB) system because the transformation of the experimental data into the center-of-mass (CM) system is not suitable, for reasons discussed by BERGMANN et al. [4.12]. Because the heavy Na_2-molecule is detected, the range of LAB angles extend from 0° to 16° only and so the CM angular resolution is very unfavorable. Generally, two CM angular intervals with $\Delta\theta \simeq 15^\circ$ contribute to a particular LAB angle. Experimental cross sections have been measured for a fixed final rotational state $j_2 = 28$ and for various initial states $j_1 \leq 12$.

Theoretical and experimental LAB cross sections versus relative energy-transfer $\Delta E/E$ (lower scale) or initial state j_1 (upper scale) are compared in Fig. 4.17 for two LAB angles. Each $\Delta j = j_2 - j_1$ distribution shows a relatively unstructured plateau for small transitions and a maximum at higher transitions beyond which the cross section rapidly declines. The relationship between these distributions, i.e., high initial states and a fixed final state, and those with $j_1 = 0$ and j_2 varied, was discussed by SCHINKE et al. [4.51] exploiting the factorization formula (4.18). According to this discussion the theoretical maximum at $j_1 = 16$ (22) for $\vartheta_{LAB} = 12.5^\circ$ (8°) corresponds to the rotational rainbow maximum.

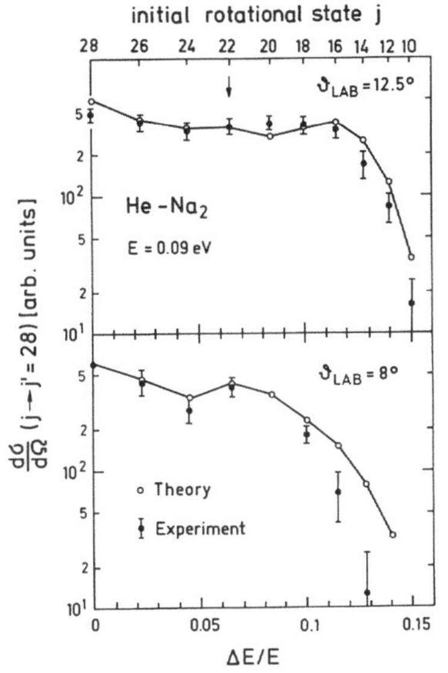

Fig.4.17. Comparison between theoretical and experimental He-Na_2 differential cross sections for $j_1 \rightarrow j_2 = 28$ transitions at laboratory angles $\vartheta_{LAB} = 12.5°$ and $8°$ versus relative energy transfer $\Delta E/E$ (lower scale) or j_1 (upper scale). The collision energy is 90 meV. The normalization is performed at $\vartheta_{LAB} = 12.5°$ and $j_1 = 22$ as indicated by the arrow (adapted from SCHINKE et al. [4.51])

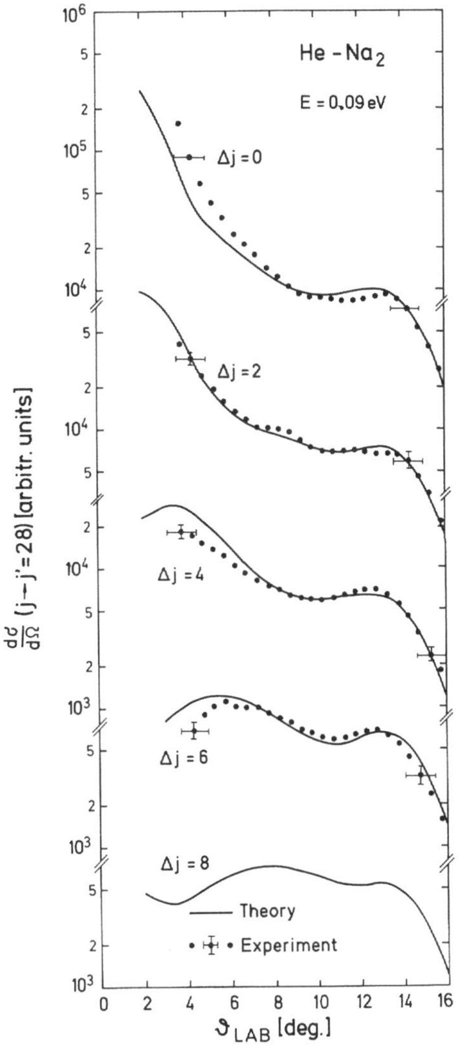

Fig.4.18. Comparison between theoretical and experimental He-Na_2 cross sections for $j_1 \rightarrow j_2 = 28$ transitions with $\Delta j = j_2 - j_1 = 0$, 2, 4, 6 and 8. Each set of cross sections is normalized separately at $\vartheta_{LAB} \simeq 12.5°$. The collision energy is E = 90 mEv (adapted from SCHINKE et al. [4.51])

Transitions with smaller or larger energy transfer are classically allowed or forbidden, respectively. The latter is clearly implied by the exponential decrease for large $\Delta E/E$. Because of the high initial states and the appreciable angular averaging, all supernumerary structures in the classically accessible region are almost completely smeared out [4.51]. The agreement between theory and experiment is satisfactory and proves the applicability of the IOS approximation as well as the accuracy of the *ab initio* potential energy surface. Disagreements are mainly observed for larger transitions, especially at the smaller scattering angle, where, compared to the experiment data, the theoretical cross sections decline too slowly into the classically forbidden region.

The same behavior is also observed in the angular distributions of single rotational transitions $j_1 \rightarrow j_2 = 28$ with $\Delta j \leq 8$ which are shown in Fig.4.18. The maximum at $\vartheta_{LAB} = 13^o$ for each set is *not* caused by a dynamical effect but is the consequence of the unfavorable CM angular resolution: all CM angles $\theta \gtrsim 90^o$ contribute to this particular LAB angle [4.12]. Although significantly obscured by the appreciable angular averaging, the rotational rainbow maxima are clearly seen for the $\Delta j = 4$, 6 and 8 cross section curves and the general predictions of Sects.4.3.4, 4.4.1 are noted. Thus, the rainbow maximum shifts to larger scattering angles with increasing Δj. Simultaneously, it becomes broader and its magnitude decreases with respect to the backward intensity. For transitions with $\Delta j \geq 8$, the kinematic peak at $\vartheta_{LAB} \simeq 13^o$ dominates all the dynamical features and rotational rainbow maxima are hardly to be seen [4.51]. The considerable disagreement between the theroetical and experimental $\Delta j = 0$ ($j_1 = 28 \rightarrow 28$) cross section is due to an experimental artifact, which fortunately does not affect the inelastic cross sections [4.12]. Again, because of the unfavorable angular averaging *and* scattering out of high rotational states, supernumerary rotational rainbows are neither theoretically nor experimentally observed.

In accordance with the Δj-distributions in Fig.4.17, theory predicts the rainbow maxima in the angular distributions at scattering angles which are slightly too small compared to the experimental results. In general, several approximations of the theoretical calculation may be responsible for this deficiency: (i) the IOS approximation, (ii) the rigid-rotor model, and (iii) the potential surface. With respect to Figs.4.11,12 and the discussion following these figures, (i) can be excluded because the energy sudden condition is satisfied for those transitions considered in Figs.4.17,18. In addition, the centrifugal sudden condition which is made in *both* approximations is believed to be valid because of the very shallow potential well and primarily wide angle scattering. A more detailed discussion, but for Ne-Na_2, has been given recently [4.16]. Full vibrating rotor calculations [4.64] using an extended He-Na_2 potential surface, which allows the inclusion of the vibrational degree of freedom, proves that the rigid-rotor approximation is very accurate for the He-Na_2 system, at least for the experimental energy of 90 meV. Thus, one could speculate that the *ab initio* calculation slightly overestimates the anisotropy of the potential energy surface and so leads to the disagreement observed in Figs.4.17,18. After all, it is not clear to us which theoretical or experimental defect might cause the disagreement.

<u>Ne-Na_2</u>

Well-resolved rotational rainbow maxima in angular distributions of $j_1 \rightarrow j_2 = 28$ differential cross sections have been reported for Ne-Na_2 [4.13]. Similar results are obtained by SERRI et al. [4.15] for Ar-Na_2 and $j_1 = 7 \rightarrow j_2$ transitions. Both experiments show nicely the increase of the rainbow angle $\theta_{max}(\Delta j)$ with the in-

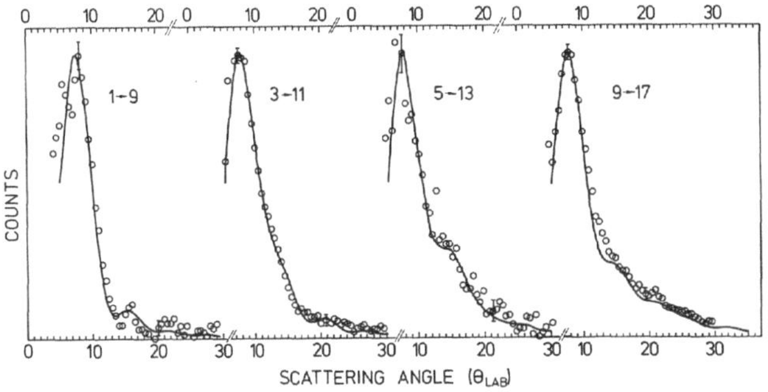

Fig.4.19. Comparison between theoretical (——) and experimental (ooo) Ne-Na_2 differential cross sections in the laboratory system for various $j_1 \to j_2$ transitions with $\Delta j = j_2 - j_1 = 8$. All curves are individually normalized to the same height at the main rainbow peak (adapted from HEFTER et al. [4.14])

elasticity Δj as predicted by the classical limit IOS approximation as well as the even simpler classical theories of ellipsoidal hard shell scattering [see (4.51) and Sect.4.6.1]. Primarily because of the high initial states, supernumerary rotational rainbows have not been resolved in these experiments. Only very recently HEFTER et al. [4.14] succeeded for the first time to resolve supernumerary rotational rainbow maxima in $j_1 = 0 \to 2$, 4 and 6 differential cross sections for Ne-Na_2. The results are shown in Fig.4.1 of the Sect.4.1.1 together with IOS calculations based on an *ab initio* potential energy surface calculated with the same accuracy as for He-Na_2 [4.16]. The agreement is striking and we note that for these transitions CS calculations give cross sections which are in excellent agreement with the IOS ones [4.16]. The primary rainbow maxima for $0 \to 2$ and 4 are off scale. Two supernumeraries are resolved for each transition. Additional maxima at larger scattering angles, which are predicted by the calculation (see Fig. 1 of [4.14]), are not sufficiently intense to be resolved. This experiment is the first real proof of the existence of rotational rainbow oscillations which, in turn, are the consequence of the quantum mechanical superposition principle. It should be underlined that they were first predicted by theory [4.25].

The influence of initial rotation on the cross section behavior is illustrated in Fig.4.19 showing experimental and IOS differential cross sections versus LAB scattering angle for equivalent angular momentum transfer $\Delta j = 8$, but for different initial states $j_1 = 1$, 3, 5 and 9. As discussed in Sect.4.4.2, the oscillation structure is rapidly quenched as j_1 is increased. Secondly, we observe that the rotational rainbow angle θ_{max} is almost independent of the initial state j_1, although the energy transfer ΔE is quite different for the various transitions considered in Fig.4.19 ($\Delta E \simeq 1.7$ meV for $1 \to 9$ and $\Delta E \simeq 4.5$ meV for $9 \to 17$). This observation indicates that angular momentum rather than energy transfer determines the Ne-Na_2 cross section behavior. It follows immediately from the factorization formula (4.18) [4.16] and is expected to be valid if the energy sudden condition is fulfilled. As in Fig.4.18 the agreement between experimental and IOS cross sections is excellent and proves the high accuracy of both the scattering calculation *and* the potential energy surface. A thorough theoretical investigation of rotational rainbow structures for Ne-Na_2 scattering is given by SCHINKE et al. [4.16]. Further comparisons between theory and experiment including many more transitions were discussed by JONES et al. [4.65]. There, also the sensitivity of the cross sections with respect to changes of the potential energy surface is investigated and thus it is estimated to which extend the *ab initio* surface is fixed by the measurements.

4.6 Approximate Analytical Expressions for $\chi(\ell,\gamma)$, $J(\ell,\gamma)$ and the Rainbow Curve $j_R(\theta)$

In this section we consider two approaches to obtain approximate, analytical expressions for $\chi(\ell,\gamma)$, $J(\ell,\gamma)$ and the rainbow curve $j_R(\theta)$, all of which have been defined and illustrated in Sects.4.3,4.4. These results are useful in several respects. First, they provide additional insight into the dependence of rotational rainbows on the collision parameters, e.g., masses, collision energy, potential parameters, and second, they can be used to approximately interpret the results of experiments which measure these rainbows.

First, we consider approximate potentials for which the JWKB phase shift, $\eta(\ell,\gamma)$, is known analytically and second we consider realistic potentials for which an approximate $\eta(\ell,\gamma)$ can be obtained analytically.

4.6.1 Hard Shell Scattering

The model of scattering of a structureless atom from a hard, anisotropic, potential shell with symmetry about the shell axis [4.66] provides a very simple understanding of rotational rainbow scattering in the strong coupling limit (impulsive collisions). Within this model the potential energy surface is approximated by $V(R,\gamma) = 0$ for $R \geq \hat{R}(\gamma)$ and $V(R,\gamma) = \infty$ for $R \leq \hat{R}(\gamma)$ where $\hat{R}(\gamma)$ describes the shape of the hard potential shell. The advantage of this model, especially within the IOS approximation, is that the various cross sections can be calculated nearly analytically. Hard shell scattering has been discussed in the context of exactly three [4.41] and two [4.52,69] dimensional classical mechanics and recently within the quantal and semiclassical IOS approximations [4.53].

Hard shell scattering is very conveniently analysed within the classical limit of the IOS approximation [4.53]. The γ-dependent JWKB phase shift (4.4) is analytically given by [4.67]

$$\eta(\ell,\gamma) = -k\hat{R}(\gamma)[(1-\beta^2)^{\frac{1}{2}} - \beta\cos^{-1}\beta] - \frac{\pi}{4} \quad , \tag{4.57}$$

with

$$\beta(\ell,\gamma) = (\ell + \frac{1}{2})(k\hat{R})^{-1} \tag{4.58}$$

and k is the wave number $k = (2\mu E)^{\frac{1}{2}}/\hbar$. Equation (4.57) is the JWKB phase shift for a hard sphere with radius $\hat{R}(\gamma)$, i.e., the sphere radius is γ-dependent. It breaks down for $\beta > 1$ but has the advantage that the deflection function $\chi(\ell,\gamma)$ (4.25) and the final action function $J(\ell,\gamma)$ (4.26) can be calculated analytically. They are

$$\chi(\ell,\gamma) = 2\cos^{-1}\beta \tag{4.59}$$

$$J(\ell,\gamma) = -2k(1-\beta^2)^{\frac{1}{2}} d\hat{R}(\gamma)/d\gamma \quad . \tag{4.60}$$

In the following we will not distinguish between j_R and $J_R = j_R + 1/2$ [see (4.26); throughout this section we assume $j_2 \gg 1/2$]. An improved JWKB phase shift for hard shell potentials is given by KORSCH [4.68] and utilized by KORSCH and SCHINKE [4.53].

Now the determinant in (4.27) is simply given by

$$D(\ell,\gamma) = 4\hat{R}(\gamma)^{-1} d^2\hat{R}(\gamma)/d\gamma^2 \tag{4.61}$$

and is *independent* of ℓ. The rainbow singularities occur whenever (4.61) becomes zero, i.e.,

$$d^2\hat{R}(\gamma)/d\gamma^2 \Big|_{\gamma=\gamma_R} = 0 \quad . \tag{4.62}$$

Thus, they are exclusively determined by the *points of inflection* $\gamma_R^{(i)}$ of the shell contour $\hat{R}(\gamma)$. This was first pointed out by BECK et al. [4.41] and they introduced the term "bulge"-effect to describe the rainbow maxima in Δj distributions. Since the rainbow orientation angles $\gamma_R^{(i)}$ $(i = 1,2,\ldots)$ in (4.62) are independent of ℓ, one obtains a very simple relationship between the rotational rainbow states $j_R^{(i)}$ and the scattering angle θ, i.e.,

$$j_R^{(i)}(\theta) = j_R^{(i)}(\pi)\, \sin\theta/2 \tag{4.63}$$

where

$$j_R^{(i)}(\pi) = -2k\, d\hat{R}(\gamma)/d\gamma \Big|_{\gamma=\gamma_R^{(i)}} \tag{4.64}$$

are the rainbow states in backward direction, i.e., $\theta = \pi$. For "sufficiently smooth" potential shells, generally *two* points of inflection $\gamma_R^{(1)}$ and $\gamma_R^{(2)}$ exist which cause the two rotational rainbow singularities $j_R^{(1)}$ and $j_R^{(2)}$ predicted in Sects.4.3.4 and 4.4.2. In the case of homonuclear molecules, $\hat{R}(\gamma)$ is symmetric about $\pi/2$ such that $\gamma_R^{(1)} - \gamma_R^{(2)}$ and the two rainbow singularities coincide. It is interesting to note that (4.63) predicts the same $\sin\theta/2$ dependence for each rainbow branch. This has been found empirically in a model study using soft potentials [4.29] (see also the inset of Fig.4.13). Equation (4.63) is easily inverted to obtain $\theta_R^{(i)}(j_2)$ which gives the angular position of the rainbow singularity for a fixed j_2. One immediately sees that $j_R^{(i)}$ is an increasing function of θ and likewise $\theta_R^{(i)}$ is an increasing function of j_2.

For fixed scattering angle, j_R depends only on the form of the hard potential shell and linearily on the wave number. The latter result has likewise been observed for the quantal rainbow maxima probing realistic (i.e., soft) potential surfaces [4.29]. We also note that an analytical expression similar to (4.64) has been derived [4.29] for a potential surface with an $\exp(-\alpha R)$ radial dependence. The importance of (4.63,64) stems from the simplicity with which the rainbow singularities are related to the collision energy E and the reduced mass μ. The energy dependence has already been discussed in Sect.4.4.1. The variation with μ was considered by SCHINKE et al. [4.16] comparing He-Na_2 and Ne-Na_2 scattering. Recalling that the two respective surfaces are strikingly similar for energies below $E \simeq 100$ meV or so [4.16], we concluded that the drastic differences in the scattering results [4.12,13] are mainly due to the different reduced masses. The ratio $\mu^{1/2}_{Ne-Na_2}$ to $\mu^{1/2}_{He-Na_2}$ is 1.9. This implies that for fixed energy and scattering angle the rainbow state j_R in the case of Ne-Na_2 is approximately twice that value for He-Na_2. This is approximately what one finds for j_{max}, however, when Fig.4.11 is compared with [Ref.4.16, Fig.5].

Because the deflection functions $\chi(\ell,\gamma)$ and $J(\ell,\gamma)$ are given analytically, the model of hard shell scattering is ideally suited to discussing the validity of the approximation (4.28). Remember, our classification of rainbow singularities in differential cross sections was based on this approximate determinant. It is shown that the nondiagonal term in (4.27), i.e., $(\partial\chi/\partial\gamma)_\ell(\partial J/\partial\ell)_\gamma$, is *exactly* zero for backward scattering, $\theta = \pi$ and $\ell = 0$. Although its magnitude rises as ℓ increases (θ decreases), (4.28) is generally a good approximation except for very low scattering angles when β in (4.58) approaches one. Its range of validity increases with the wave number k and the extent of the potential shell, i.e., $\hat{R}(\gamma)$. Such behavior was anticipated in Sect.4.3.2.

A frequently used potential shell is the ellipse with major and minor semiaxes A and B, i.e.,

$$\hat{R}(\gamma) = B(1 - e^2 \cos^2\gamma)^{-1/2} \tag{4.65}$$

with eccentricity

$$e = (A^2 - B^2)^{1/2}/A \quad . \tag{4.66}$$

Several approximate cross section expressions valid for $e^2 \ll 1$ are given [4.53]. A very useful expression for the classical rainbow curve has been derived by BOSANAC [4.52] using classical mechanics in two dimensions. It is

$$j_R(\theta) = 2k(A - B) \sin\theta/2 \tag{4.67}$$

and is valid for $\mu/I \ll 1$, where I is the moment of inertia of the ellipsoid. The IOS result (4.63,64) agrees with (4.67) up to fourth order in e [4.53]. The analogue of (4.67) for off-center hard ellipsoid scattering was derived by BOSANAC and BUCK [4.69].

BECK and coworkers [4.9,10,41] used the reduced recoil velocity u^* defined in (4.55) as the independent variable rather than the rotational momentum transfer Δj. The relationship between u^* and j_2 is given by (4.56) valid for $j_2 \gg 1$. Their analysis of the rainbow singularities in the exact, three-dimensional classical cross section is based on the vanishing of the determinant $\partial(\theta,u^*)/\partial(b,\bar{\gamma}_1)$, with b the impact parameter and $\bar{\gamma}_1$ the initial orientation angle of the molecule with respect to the initial relative momentum. The rainbow singularities u_R^* in the u^*-distribution depend on the scattering angle θ according to

$$u_R^*(\theta) = [(1 - Q^2 \sin^2\theta)^{\frac{1}{2}} + Q \cos\theta](1 + Q)^{-1} \tag{4.68}$$

which, as (4.63,64) for the IOS approximation is valid for arbitrary potential shells. In (4.68) Q depends only on the ratio μ/I and the shape of the potential shell. For ellipsoid scattering the parameter Q can be given approximately [4.9], i.e.,

$$Q^{(i)} \simeq \frac{\mu}{I} [A - B \mp Z(1 + B/A)^{-\frac{1}{2}}]^2 \quad . \tag{4.69}$$

Z measures the shift between the center-of-mass and the center-of-symmetry of the ellipsoid. It is zero for homonuclear molecules. Equation (4.69) is valid for small Z and is exact for Z = 0. In accordance with the predictions of the classical limit IOS approximation in Sect.4.3.4, two rotational rainbow singularities $\mu_R^{*(1)}$ and $u_R^{*(2)}$ exist for scattering with heteronuclear molecules. The separation between $u_R^{*(2)}$ and $u_R^{*(1)}$ becomes successively smaller as Z is decreased and zero for Z = 0, i.e., for scattering with a homonuclear molecule. For fixed parameters A, B, and Z, the rotational rainbow singularities $u_R^{*(i)}$ are *independent* of the collision energy. Thus, recalling (4.56), the analysis of BECK et al. [4.41] predicts the same $E^{\frac{1}{2}}$-dependence of $j_R(\theta)$ as (4.64,67). Obviously, the classical treatment does not describe those structures which are due to quantum mechanical interference, i.e., supernumerary rotational rainbows, for example.

The hard shell model offers a very nice and easy interpretation of rotational rainbow scattering because it relates the rainbow singularities directly to the shape of the potential shell. Equations (4.67-69) are, in particular, of considerable practical importance because they allow for a *crude* estimation of the potential anisotropy, i.e., (A - B) from measured cross sections [4.9,17,69]. On the other hand, if the potential surface is known, the hard shell model predicts the dependence of the rainbow curve on the collision energy and the masses of the scattering partners in good accord with quantal studies using realistic potential energy surfaces [4.16,29]. However, we stress that because of the unrealistic radial dependence, the hard shell model fails drastically to describe details of the cross sections. This has been discussed for differential [4.53] as well as for integral cross sections [4.53,70].

Next, we consider realistic potentials for which approximate, high energy phase shifts are known analytically.

4.6.2 High Energy Approximation

Here we consider several realistic forms for the interaction potential $V(R,\gamma)$, but where we obtain approximate, analytical expressions for $J(\ell,\gamma)$, $\chi(\ell,\gamma)$ and $j_R(\theta)$. The high energy, Eikonal approximation to the phase shift is the basis for our approach. Recall that in this approximation [4.71]

$$\eta(\ell,\gamma) = -(2v)^{-1} \int_{-\infty}^{\infty} dz\ V(R(z,\ell),\gamma) \tag{4.70}$$

where

$$R^2(z,\ell) = z^2 + b^2 \quad ; \quad \ell = \mu v b \quad . \tag{4.71}$$

Atomic units ($\hbar = 1$) will be used in this section. In (4.71) μ is the collision system reduced mass, b is the impact parameter, v is the initial relative velocity, and k is the wave number. Equation (4.70) is equivalently rewritten as

$$\eta(\ell,\gamma) = -\frac{\mu}{k} \int_{\ell/k}^{\infty} dR\ V(R,\gamma)\left(1 - \frac{\ell^2}{k^2R^2}\right)^{-\frac{1}{2}} \quad . \tag{4.72}$$

We first consider the potential energy surface given in (4.52) with an inverse power R-dependence, i.e.,

$$V(R,\gamma) = g(\gamma)R^{-n} \quad . \tag{4.73}$$

For this potential (4.72) can be evaluated analytically [Ref.4.71, Eq. (4.18)] to give

$$\eta(\ell,\gamma) = -\mu f(n) g(\gamma) k^{n-2} \ell^{1-n} \tag{4.74}$$

with f(n) defined as [Ref.4.71, Eq. (4.19)]

$$f(n) = \frac{\pi^{\frac{1}{2}}}{2}\,\Gamma\!\left(\frac{n-1}{2}\right)/\Gamma\!\left(\frac{n}{2}\right) \tag{4.75}$$

and Γ is the Gamma function [Ref.4.35, Chap.6]. Inserting (4.74) into (4.25,26) for $\chi(\ell,\gamma)$ and $J(\ell,\gamma)$, respectively, we find

$$\chi(\ell,\gamma) = 2\mu(n-1)f(n)g(\gamma)k^{n-2}\ell^{-n} \tag{4.76}$$

$$J(\ell,\gamma) = -2\mu f(n)g'(\gamma)k^{n-2}\ell^{1-n} \tag{4.77}$$

with $g'(\gamma) = dg(\gamma)/d\gamma$.

Recall that classical rainbows are given by $D(\ell,\gamma) = 0$ with the determinant defined in (4.27) and, approximately, in (4.28). In the present case it turns out that either of the above equations gives a root γ_R which is independent of ℓ. Thus, just as for hard shell potentials, we obtain a very simple relationship between the rotational rainbow state j_R and the scattering angle. Solving (4.76) for ℓ, inserting into (4.77) and replacing χ by θ [which is possible because χ in (4.76) is a monotonically decreasing function of ℓ], we finally obtain

$$j_R(\theta) = dk^{(n-2)/n}\ \theta^{(n-1)/n} \tag{4.78}$$

where the constant d depends mainly on the potential energy surface and only very slightly on the reduced mass. It is

$$d = -g'(\gamma_R)[(n-1)g(\gamma_R)]^{-1}[2\mu(n-1)f(n)g(\gamma_R)]^{1/n} \quad . \tag{4.79}$$

Equation (4.78) can be compared to the result for the hard shell model (using the JWKB phase shift), (4.63,64) and where qualitative agreement is seen, i.e., j_R increases with both θ and k. Replacing $(n - 2)/n$ by one, valid for large n, (4.78) predicts for fixed θ a linear dependence on k. Note, such behavior has been surmised empirically [4.16] for potential surfaces of the form (4.52) and it is now substantiated by the present Eikonal analysis using (4.73). It is trivial to invert (4.78) to obtain the angle $\theta_R(j_2)$ where a classical rainbow occurs, i.e.,

$$\theta_R(j_2) = d^{-n/(n-1)}\ k^{-(n-2)/(n-1)}\ j_2^{n/(n-1)} \quad . \tag{4.80}$$

This shows that θ_R is an *increasing* function of j_2 for fixed k and a *decreasing* function of k for fixed j_2.

Next, we again consider the model He-(homonuclear)CO system described in Sect. 4.4.1 to see how well the correct classical IOS rainbow curve shown in Fig.4.8 is approximated by the forms given in (4.63,80), respectively. Note the rainbow curve in Fig.4.8 is based on the full potential in (4.49) which consists of a repulsive and an attractive part, while (4.80) is derived for a potential with a repulsive term only, i.e., (4.73). However, it is a reasonable strategy to neglect the attractive part for a collision energy of $E = 150\varepsilon$. We have taken the correct values θ_R and $\sin\theta_R/2$ and divided them by $j_2^{12/11}$ $(n = 12)$ and j_2, respectively, and in Table 4.2 we present the results. Overall, both functional forms describe the j_2-dependence of the correct θ_R quite well and with about equal accuracy.

As seen in Table 4.2 the functional form given by (4.80) for $n = 12$ reproduces the j_2-dependence of the correct classical rainbow curve θ_R; however, as seen in Fig.4.8 the correct quantal (and uniform semiclassical) rainbow maxima θ_{max} are considerably shifted to greater scattering angles. This shift in θ is a result of the shift in j_R discussed in Sect.4.3.4. An expression for the shift $j_R - j_{max}$ at a fixed θ was given in (4.41,42), based on the uniform Airy analysis of the classical limit IOS quantal scattering amplitude. If we regarded the quantal curve in

Table 4.2. Predicted j_2-dependence of θ_R as derived from the Eikonal phase shift approximation and the hard shell model[a]

j_2	θ_R	$j_2^{-12/11}\theta_R$	$j_2^{-1}\sin\theta_R/2$
2	19[b]	8.9[c]	8.3(-2)[d]
4	32	7.1	6.9(-2)
6	43	6.1	6.1(-2)
8	60	6.2	6.2(-2)
10	74	6.0	6.0(-2)
12	95	6.3	6.1(-2)
14	115	6.5	6.0(-2)
16	156	7.6	6.1(-2)

[a] Potential and collision parameters are those for He-N_2 (Sect.4.4.1)
[b] Correct classical limit IOS result
[c] Approximation based on Eikonal phase shift, i.e., (4.80)
[d] Approximation based on hard shell model, i.e., (4.63). Numbers in paranthesis indicate powers of ten.

Fig.4.8 as the "experimental" data, then we would have to proceed with much caution in trying to use (4.63,80) to fit and interpret the experimental results. Because such an approach is so appealing, let us proceed to improve the classical rainbow curve (4.80) by including the shift $j_R - j_{max}$.

Using (4.41,42) and for the potential given in (4.73) it is easy to show that

$$j_{max}(\theta) = s\theta^{(n-1)/n} - t\theta^{(n-1)/3n} \tag{4.81}$$

where s and t are positive constants which are obtained from $g(\gamma)$ in (4.73). Comparing (4.81) with (4.78) we see that j_{max} is less than j_R at a given value of θ. Qualitatively, that is what is observed in Fig.4.8. Unfortunately, we cannot easily determine whether the differences between the j_R and j_{max} curves shown in that figure are well described by the differences between (4.81) and (4.78) until we at least know the ratio t/s. For the model He-N_2 system we have been considering, s can be determined empirically from Table 4.2. The constant t is determined from s and $V(R,\gamma)$ given by (4.73). The result is

$$j_{max}(\theta) = 0.18\theta^{11/12} - 0.72\theta^{11/36} \quad . \tag{4.82}$$

To see how well this expression fits the quantal sudden "data", we plot in Fig. 4.20 $j_{max}(\theta)$ given above, along with the quantal sudden $j_{max}(\theta)$ (already plotted in Fig.4.8), and for reference we plot $j_R(\theta)$ given by (4.82) with the $\theta^{11/36}$ term deleted. Clearly, $j_R(\theta)$ is an inadequate quantitative representation of $j_{max}(\theta)$; however, (4.82) fits the quantum "data" quite well.

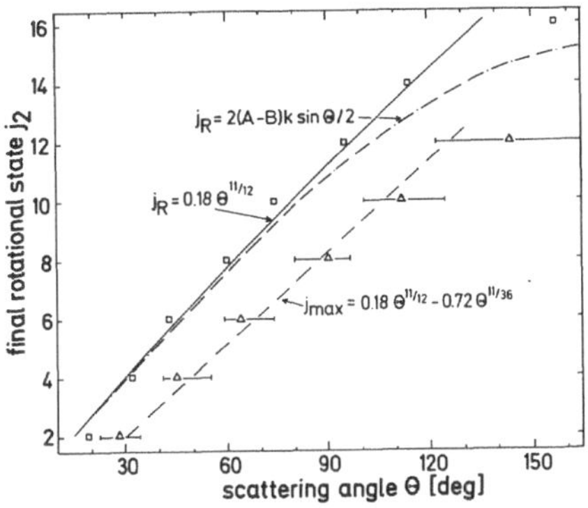

Fig.4.20. Exact classical (□) and quantal (Δ) rainbow curves $j_R(\theta)$ and $j_{max}(\theta)$ for the model He-N_2 system defined in Sect.4.4.1 at $E = 150\varepsilon$. Comparison with the quantal corrected Eikonal result (4.82) (---), the classical Eikonal result, i.e., (4.81) with the second term deleted (——), and the hard ellipsoid result (4.67) (·-·-·-·)

It thus appears that the classical rainbow curve should not be used to fit experimental rainbow maxima with an assumed potential. It is essential that the shift in the classical rainbow curve be included in any attempt to fit experimental data. We recommend incorporating the results of a uniform semiclassical analysis, i.e., (4.41,42), which account for the shift j_{max} away from j_R, to obtain a rainbow curve $j_{max}(\theta)$ to compare with or fit experimental data. We should stress that the quantities s and t appearing in (4.81) are not to be regarded as independent of each other. They are both functions of the parameters which appear in $g(\gamma)$ in (4.73).

Quite significant deviations between the exact IOS rainbow curves $j_{max}(\theta)$ and $j_R(\theta)$ and those obtained from the Eikonal analysis, i.e., (4.82) and $j_R = 0.18\ \theta^{11/12}$, are observed for large angles $\theta \gtrsim 120^\circ$. This is expected because the Eikonal approximation is a small angle approximation. The action integral in (4.70) is evaluated along straight-line "trajectories" $R(t)$, i.e., (4.71). Also shown in Fig. 4.20 is the classical rainbow curve (4.67) obtained from the hard ellipsoid model with A-B = 0.564 bohr. The semiaxes A and B are defined as the classical turning points for $\ell = 0$ and $\gamma = 0^\circ$ and $\gamma = 90^\circ$, respectively [4.53]. At small angles ($\theta \lesssim 90^\circ$) it fits the exact classical rainbow points as well as the Eikonal expression and is more accurate for larger θ. It thus appears that the hard shell model gives a better overall description of the classical rainbow curve. However, we concede that the definition of the semiaxes is quite arbitrary. As in the case of hard shell scattering [4.53], the good description of the rainbow curve by (4.82) *does not* imply that the individual cross sections calculated with the approximate phase (4.70) are quantitatively accurate.

We conclude this section by briefly tabulating in Table 4.3 the high energy phase shifts for several potentials where the γ-dependence is contained in the parameters of the potential directly. These results are all simple generalizations of ones well known from potential scattering [4.30]. By differentiating $2\eta(\ell,\gamma)$ given there with respect to ℓ and γ, the deflection functions $\chi(\ell,\gamma)$ and $J(\ell,\gamma)$, respectively, are obtained. For simplicity, only the potential parameters ε and R_m are considered to be functions of γ.

Table 4.3. High energy phase shifts $\eta(\ell,\gamma)$ for three potentials $V(R,\gamma)$

a) Lennard-Jones n - 6

$$V(R,\gamma) = \frac{6\varepsilon(\gamma)}{n-6}\left[\left(\frac{R_m(\gamma)}{R}\right)^n - \frac{n}{6}\left(\frac{R_m(\gamma)}{R}\right)^6\right]$$

$$\eta(\ell,\gamma) = \frac{\pi\mu R_m(\gamma)\varepsilon(\gamma)}{k}\left[\frac{3}{16}\frac{n}{n-6}\left(\frac{kR_m(\gamma)}{\ell}\right)^5 - \frac{6f(n)}{\pi(n-6)}\left(\frac{kR_m(\gamma)}{\ell}\right)^{11}\right]$$

$$f(n) = \frac{\pi^{\frac{1}{2}}}{2}\,\Gamma\left(\frac{n-1}{2}\right)/\Gamma\left(\frac{n}{2}\right)$$

b) Buckingham-Slater

$$V(R,\gamma) = \frac{6\varepsilon(\gamma)}{\alpha-6}\left\{\exp[\alpha(1 - R/R_m(\gamma))] - \frac{\alpha}{6}\left(\frac{R_m(\gamma)}{R}\right)^6\right\}$$

$$\eta(\ell,\gamma) = \frac{\mu R_m(\gamma)\varepsilon(\gamma)}{k(\alpha-6)}\left(\frac{3\pi\alpha}{16}\left[\frac{kR_m(\gamma)}{\ell}\right]^5 - \left(\frac{\pi}{\alpha}\right)^{\frac{1}{2}}\left[\frac{\ell}{kR_m(\gamma)}\right]^{\frac{1}{2}}\exp\{\alpha[1 - \ell/kR_m(\gamma)]\}\right)$$

c) Morse

$$V(R,\gamma) = \varepsilon(\gamma)\Big(\exp\{2\alpha[1 - R/R_m(\gamma)]\} - 2\exp\{\alpha[1 - R/R_m(\gamma)]\}\Big)$$

$$\eta(\ell,\gamma) = \left(\frac{2\pi}{\alpha}\right)^{\frac{1}{2}}\mu\left(\frac{\ell R_m(\gamma)}{k}\right)^{\frac{1}{2}}\varepsilon(\gamma)k^{-\frac{1}{2}}\exp\{\alpha[1 - \ell/kR_m(\gamma)]\}\Big(1 - 2^{-3/2}\exp\{\alpha[1 - \ell/kR_m(\gamma)]\}\Big)$$

4.7 Conclusion and Prognosis

Rotational rainbow features are expected to dominate rotationally inelastic differential cross sections for *any* atom-diatom system, provided the collision is mainly impulsive and many rotational states are energetically open. All structures predicted by the ultra simple IOS approximation, i.e., main rainbow maxima and supernumerary rotational rainbows in the classically allowed Δj or θ-region, are now confirmed by state-to-state experiments. They are easily and elegantly interpreted in terms of the classical limit of the IOS scattering amplitude. First quantitative comparisons between theory and experiment using accurate *ab initio* potential surfaces have been reported (and partly discussed in this chapter) for He, Ne-Na_2 and gave excellent agreement.

Rotational rainbow structures are extremely sensitive to the anisotropy of the potential energy surface at small internuclear distances. Thus, direct inversion of experimental cross sections (if possible) will yield valuable information on the interaction potential in this region. An inversion scheme which is based on the IOS approximation has been suggested [4.72]. This method consists of two steps. First, the scattering phaseshift $\eta(\ell,\gamma)$ is constructed from the differential cross section data exploiting simple relations which are derived from the semiclassical version of the IOS scattering amplitude. Second, Firzov's method as known from the inversion of elastic cross sections in the case of isotropic potential scattering [4.73] is applied to construct for fixed γ the potential $V(R,\gamma)$ from the ℓ-dependence of $\eta(\ell,\gamma)$. Here, one utilizes the fact that the orientation angle γ is treated as a parameter within the IOS approximation. So far this inversion scheme has only been tested in a model study with theoretical He-Na_2 cross sections taken as quasiexperimental input data.

The entire discussion in this chapter was based on the rigid-rotor approximation, i.e., the vibrational degree of freedom of the molecule was frozen at its asymptotic equilibrium distance. The effect of molecular vibration on the rotational rainbow structures is investigated for He-Na_2 [4.64]. The main results of this study can be summarized as follows. For the collision energies considered ($E \leq 150$ meV), the probability of vibrational excitation is extremely small and the inclusion of the vibrational degree of freedom has only a negligible effect on the rotational rainbow structures in the vibrationally elastic cross sections. The vibrationally inelastic cross sections show rotational rainbow features in the same way as the vibrationally elastic ones. Surprisingly, it is found that the rainbow angle $\theta_{max}(\Delta j)$ for a fixed rotational transition is almost independent of whether $\Delta n = 0$ or $\Delta n = 1$. This is in qualitative accord with recently reported state-to-state cross sections for Ar- Na_2 [4.74].

Structures in the energy dependence of rotationally inelastic integral cross sections were discussed by SCHINKE [4.75,76]. They are closely related to rotational rainbows in the corresponding differential cross sections. The structures are (i) dynamical thresholds at low energies and (ii) quantum undulations at higher energies. The latter stem from the energy dependence of the supernumerary rainbow maxima and therefore they are regarded as the analogue of the glory maxima in the case of isotropic potential scattering [4.30,73]. The existence of dynamical constraints on the transfer of angular momentum in rotationally inelastic collisions is recently confirmed experimentally by DEXHEIMER et al. [4.77] for the systems He, Xe-I_2. Interference oscillations in rotationally inelastic integral cross sections for the Ar-NO collision system are resolved by ANDRESEN et al. [4.78]. These structures are due to the asymmetry of the heteronuclear target molecule as discussed in Sect.3.4 and Fig.4.5 for the differential cross section. Structural changes of rotational rainbows within various classical treatments are investigated by KORSCH and RICHARDS [4.79] and KORSCH and POPPE [4.80].

Throughout this chapter we assumed that either the potential energy surface was purely repulsive or that the depth of the van der Waals minimum was extremely small compared to the collision energy. It is this assumption which actually allowed us to classify the pronounced structures in the rotationally inelastic cross sections as rotational rainbows. Preliminary calculations using a potential surface with an appreciably deep well showed that in this case a classification into different rainbow types is in general not possible [4.81]. This was previously noted by THOMAS [4.40] and BOWMAN and LEE [4.33]. In addition, the topology of the rainbow singularities becomes rapidly more complicated as the well depth is increased. These more general cases should be investigated in more detail in the future. A first step in this direction is discussed by BENTLEY [4.82]. However, we recall that a rigorous theoretical study is hindered when the potential has a strong attraction at large radial distances because then the centrifugal sudden approximation breaks down.

Interesting new features related to rotational rainbows as discussed in this chapter may be found in more complex collisions, i.e., diatom-diatom or atom-triatom scattering, for example. Recently, SCHINKE [4.83] showed that rotational rainbows will be common structures in diatom-solid surface scattering provided the same conditions as in the case of gas phase collisions hold, i.e., heavy molecules and high energies. The recent experiments by KLEYN et al. [4.84] indicate the existence of rotational rainbows in molecule surface collisions. They have also been predicted by quasiclassical trajectory calculations [4.85]. Finally we note that rainbow singularities may also occur for other inelastic processes. DROLSHAGEN et al. [4.86] reported the observation of *vibrational rainbows* in a theoretical study considering Li^+-H_2 scattering at high collision energies. Using classical S-matrix theory [4.23,24] in the breathing sphere approximation, they explained maxima in the $\Delta n \neq 0$ differential cross sections as rainbow singularities arising from an extremum of the final action function n_2 versus the initial phase of the oscillator.

Acknowledgements. We are grateful to Dr. H.J. Korsch (Kaiserslautern) for calculating the classical and semiclassical cross sections in Table 4.1. We are also grateful to Dr. W. Schepper (Bielefeld) for the permission to publish the experimental K - N_2 and CO cross sections in Fig.4.15. R. Schinke acknowledges many fruitful discussions with Dr. W. Müller (Kaiserslautern) concerning the potential energy calculations for He-Na_2 and Ne-Na_2. He also acknowledges stimulating discussions with Prof. K. Bergmann (Kaiserslautern) concerning the corresponding experiments. JMB thanks the National Science Foundation for its support.

References

4.1 For a recent review of ion-molecule scattering see M. Faubel, J.P. Toennies: Adv. At. Mol. Phys. *13*, 229 (1977)
4.2 For a recent review of atom-molecule scattering see H.G. Loesch: Adv. Chem. Phys. *42*, 421 (1980)
4.3 U. Buck, F. Huisken, J. Schleusener, H. Pauly: Phys. Rev. Lett. *38*, 680 (1977);
U. Buck, F. Huisken, J. Schleusener, J. Schäfer: J. Chem. Phys. *72*, 1512 (1980)
4.4 J. Andres, U. Buck, F. Huisken, J. Schleusener, F. Torello: J. Chem. Phys. *73*, 5620 (1980)
4.5 W.R. Gentry, C.F. Giese: J. Chem. Phys. *67*, 5389 (1977)
4.6 U. Buck, F. Huisken, J. Schleusener: J. Chem. Phys. *68*, 5654 (1978);
U. Buck, F. Huisken, J. Schleusener, J. Schäfer: J. Chem. Phys. *74*, 535 (1981)
4.7 W.R. Gentry, C.F. Giese: Phys. Rev. Lett. *39*, 1259 (1977)
4.8 M. Faubel, K.H. Kohl, J.P. Toennies: J. Chem. Phys. *73*, 2506 (1980);
M. Faubel, K.H. Kohl, J.P. Toennies, K.T. Tang, Y.Y. Yung: Faraday Discussion *73* (1982)
4.9 W. Schepper, U. Ross, D. Beck: Z. Phys. A*290*, 131 (1979)
4.10 D. Beck, U. Ross, W. Schepper: Phys. Rev. A*19*, 2173 (1979)
4.11 K. Bergmann, R. Engelhardt, U. Hefter, J. Witt: J. Chem. Phys. *71*, 2726 (1979)
4.12 K. Bergmann, U. Hefter, J. Witt: J. Chem. Phys. *72*, 4777 (1980)
4.13a K. Bergmann, U. Hefter, A. Mattheus, J. Witt: Chem. Phys. Lett. *78*, 61 (1981)
4.13b P.L. Jones, E. Gottwald, U. Hefter, K. Bergmann: J. Chem. Phys. *78* (1983)
4.14 U. Hefter, P.L. Jones, A. Mattheus, J. Witt, K. Bergmann, R. Schinke: Phys. Rev. Lett. *46*, 915 (1981)
4.15 J.A. Serri, A.M. Morales, W. Moskowitz, D.E. Pritchard, C.H. Becker, J.L. Kinsey: J. Chem. Phys. *72*, 6304 (1980)
4.16 R. Schinke, W. Müller, W. Meyer: J. Chem. Phys. *76*, 895 (1982)
4.17 J. Andres, U. Buck, H. Meyer, J.M. Launay: J. Chem. Phys. *76*, 1417 (1982)
4.18 J.C. Light: "Inelastic Scattering Cross Sections I: Theory", in *Atom-Molecule Collision Theory*, ed. by R.B. Bernstein (Plenum, New York 1979) Chap.6, pp.239
4.19 D. Secrest: "Rotational Excitation I: The Quantal Treatment", in *Atom-Molecule Collision Theory*, ed. by R.B. Bernstein (Plenum, New York 1979) Chap.8, pp.265
4.20 D.J. Kouri: "Rotational Excitation II: Approximation Methods", in *Atom-Molecule Collision Theory*, ed. by R.B. Bernstein (Plenum, New York 1979) Chap.9, pp.301
4.21 R. Schinke, P. McGuire: J. Chem. Phys. *71*, 4201 (1979)
4.22 M.D. Pattengill: "Rotational Excitation III: Classical Trajectory Methods", in *Atom-Molecule Collision Theory*, ed. by R.B. Bernstein (Plenum, New York 1979) Chap.10, pp.359
4.23 W.H. Miller: J. Chem. Phys. *53*, 1949 (1970);
R.A. Marcus: J. Chem. Phys. *54*, 3965 (1971)
4.24 For reviews, see W.H. Miller: Adv. Chem. Phys. *25*, 69 (1974); Adv. Chem. Phys. *30*, 77 (1975)
4.25 R. Schinke: Chem. Phys. *34*, 65 (1978)

4.26 J.M. Bowman, K.-T. Lee: Chem. Phys. Lett. *60*, 212 (1979)
4.27 H.J. Korsch, R. Schinke: J. Chem. Phys. *73*, 1222 (1980)
4.28 J.M. Bowman: Chem. Phys. Lett. *62*, 309 (1979)
4.29 R. Schinke: J. Chem. Phys. *72*, 1120 (1980)
4.30 H. Pauly: "Elastic Scattering Cross Sections I: Spherical Potentials', in *Atom-Molecule Collision Theory*, ed. by R.B. Bernstein (Plenum, New York 1979) Chap.4, pp.111
4.31 K.W. Ford, J.A. Wheeler: Ann. Phys. *7*, 259 (1959)
4.32 See, for example, J.N.L. Connor: Mol. Phys. *25*, 181 (1973)
4.33 J.M. Bowman, K.-T. Lee: J. Chem. Phys. *74*, 2664 (1981)
4.34 C.W. McCurdy, W.H. Miller: J. Chem. Phys. *67*, 463 (1977)
4.35 A. Abramowitz, I. Stegun: *Handbook of Mathematical Functions* (Dover, New York 1965)
4.36 W.H. Miller: J. Chem. Phys. *53*, 3578 (1970)
4.37 T. Mullnoney, G.C. Schatz: Chem. Phys. *45*, 213 (1980)
4.38 For a recent review see A.P. Clark, A.S. Dickinson, D. Richards: Adv. Chem. Phys. *36*, 63 (1977)
4.39 L.D. Thomas: J. Chem. Phys. *67*, 5224 (1977)
4.40 L.D. Thomas: J. Chem. Phys. *73*, 5905 (1980)
4.41 D. Beck, U. Ross, W. Schepper: Z. Phys. A*293*, 107 (1979); Z. Phys. A*299*, 97 (1981)
4.42 R. Schinke: Chem. Phys. *47*, 287 (1980)
4.43 A.S. Dickinson: Comput. Phys. Com. *17*, 51 (1979)
4.44 S. Augustin, W.H. Miller: Chem. Phys. Lett. *28*, 149 (1974)
4.45 S. Green, P. Thaddeus: Astrophys. J. *205*, 766 (1976)
4.46 S. Chapman, S. Green: J. Chem. Phys. *67*, 2317 (1977)
4.47 S. Green: Chem. Phys. Lett. *38*, 293 (1976)
4.48 S. Green: Chem. Phys. *31*, 425 (1978)
4.49 L.D. Thomas, W.P. Kraemer, G.H.F. Diercksen: Chem. Phys. *51*, 131 (1980)
4.50 J.M. Bowman, K.T. Lee, J.G. Sachs: Chem. Phys. Lett. *74*, 90 (1980)
4.51 R. Schinke, W. Müller, W. Meyer, P. McGuire: J. Chem. Phys. *74*, 3916 (1981)
4.52 S. Bosanac: Phys. Rev. A*22*, 2617 (1980)
4.53 H.J. Korsch, R. Schinke: J. Chem. Phys. *75*, 3850 (1981)
4.54 R. Böttner, U. Ross, J.P. Toennies: J. Chem. Phys. *65*, 733 (1976)
4.55 V. Staemmler: Chem. Phys. *7*, 17 (1975); Chem. Phys. *17*, 187 (1976)
4.56 L.D. Thomas, W.P. Kraemer, G.H.F. Diercksen: Chem. Phys. *30*, 33 (1978)
4.57 L.D. Thomas, W.P. Kraemer, G.H.F. Diercksen, P. McGuire: Chem. Phys. *27*, 237 (1978)
4.58 D. Poppe, R. Böttner: Chem. Phys. *30*, 375 (1978)
4.59 G.D. Billing: Chem. Phys. *36*, 127 (1979)
4.60 D.A. Michia, E. Villalonga, J.P. Toennies: Chem. Phys. Lett. *62*, 238 (1979); L.H. Beard, D.A. Micha: J. Chem. Phys. *74*, 6700 (1981)
4.61 L.D. Thomas: "Rainbow Scattering In Inelastic Molecular Collisions", in *Potential Energy Surfaces and Dynamic Calculations*, ed. by D.G. Truhlar (Plenum, New York 1981)
4.62 W. Eastes, U. Ross, J.P. Toennies: Chem. Phys. *39*, 407 (1979)
4.63 L.D. Thomas, W.P. Kraemer, G.H.F. Diercksen: Chem. Phys. Lett. *74*, 445 (1980)
4.64 W. Müller, R. Schinke: J. Chem. Phys. *75*, 1219 (1981)
4.65 P.L. Jones, U. Hefter, A. Mattheus, K. Bergmann, W. Müller, W. Meyer, R. Schinke: Phys. Rev. A*26*, 1283 (1982)
4.66 R.A. La Budde, R.B. Bernstein: J. Chem. Phys. *55*, 5499 (1971)
4.67 R.B. Bernstein: J. Chem. Phys. *36*, 1403 (1962);
P.E. Siska: J. Chem. Phys. *59*, 3439 (1973);
M.E. Riley, P.E. Siska: J. Chem. Phys. *61*, 2435 (1974)
4.68 H.J. Korsch: J. Chem. Phys. *69*, 1311 (1978)
4.69 S. Bosanac, U. Buck: Chem. Phys. Lett. *81*, 315 (1981)
4.70 M.H. Alexander, P.J. Dagdigian: J. Chem. Phys. *73*, 1233 (1980)
4.71 See, for example, M.S. Child: *Molecular Collision Theory* (Academic, New York 1974) Chap.4
4.72 R. Schinke: J. Chem. Phys. *73*, 6117 (1980)
4.73 See, for example, U. Buck: Adv. Chem. Phys. *30*, 313 (1975)

4.74 J.A. Serri, C.H. Becker, M.B. Elbel, J.L. Kinsey, W.P. Moskowitz, D.E. Pritchard: J. Chem. Phys. *74*, 5116 (1981)
4.75 R. Schinke: J. Chem. Phys. *75*, 5449 (1981)
4.76 R. Schinke: J. Chem. Phys. *75*, 5205 (1981)
4.77 S.L. Dexheimer, M. Durand, T.A. Brunner, D.E. Pritchard: J. Chem. Phys. , 4996 (1982)
4.78 P. Andresen, H. Joswig, H. Pauly, R. Schinke: J. Chem. Phys. *77*, 2204 (1982)
4.79 H.J. Korsch, D. Richards: J. Phys. B*14*, 1973 (1981)
4.80 H.J. Korsch, D. Poppe: J. Chem. Phys. *69*, 99 (1982)
4.81 R. Schinke, H.J. Korsch, D. Poppe: J. Chem. Phys. *77* (1982)
4.82 J. Bentley: J. Chem. Phys. *73*, 4708 (1980)
4.83 R. Schinke: J. Chem. Phys. *76*, 2352 (1982)
4.84 A.W. Kleyn, A.C. Luntz, D.J. Auerbach: Phys. Rev. Lett. *75*, 1033 (1981)
4.85 J.M. Bowman, S. Park: J. Chem. Phys. *76*, 1168 (1982)
4.86 G. Drolshagen, H.R. Mayne, J.P. Toennies: J. Chem. Phys. *75*, 196 (1981)

5. Quantum Mechanical Treatment of Electronic Transitions in Atom-Molecule Collisions

M. Baer

With 16 Figures

The treatment of electronic transitions taking place during atomic collisions has been considered important since the early days of quantum mechanics. BORN and OPPENHEIMER [5.1] were the first to study this problem and their theory still serves as the starting point for any other treatment or approximation. We shall refer to this later, though not extensively.

Strictly speaking, electronic nonadiabatic transitions occur due to the breakdown of the Born-Oppenheimer (BO) approximation. In general, one may distinguish between two forms of electronic nonadiabatic transitions: (i) those originating from the radial motion which, in the atom-atom case, arises due to the translational motion and in more general cases due to vibrational and angular motions as well; (ii) those originating from the rotation of the body axis of a group of atoms with respect to an axis fixed in space.

Radial coupling was first treated by ZENER [5.2], LANDAU [5.3] and STÜCKELBERG [5.4]. They found that the BO approximation breaks down when two adiabatic molecular states of the same symmetry very closely approach each other. Two cases are to be distinguished according to the strength of the nonadiabatic coupling term. The first (treated by ZENER [5.2], LANDAU [5.3], STÜCKELBERG [5.4] and others [5.5-9]) is characterized by the fact that the two diabatic curves or surfaces intersect. The second, treated mainly by DEMKOV [5.10], is characterized by a weak nonadiabatic coupling term (for instance, spin-orbit coupling) which yields two nonintersecting diabatic curves (or surfaces).

Rotational coupling was first discussed by KRONIG [5.11] who found that in order to properly treat electrons and nuclei on the whole, one should deal with two different systems of coordinates, one fixed in space (SF) and the other fixed in the molecule (BF). He found that the transformation from SF to BF yields additional nonadiabatic couplings between various electronic states due to the conservation of total angular momentum. The nature of this coupling was studied extensively by KRONIG [5.11] and others [5.12-18].

The main difference between the two types of coupling resides in the fact that radial coupling may only cause transitions between states of the same symmetry, whereas rotational coupling can mix states of the same and different symmetry.

The literature mentioned in the previous paragraphs was concerned with either diatomic molecules or atom-(ion)-atom collisions. Electronic nonadiabatic transitions induced in three-body collisions (atom-diatom collisions) were, until very recently, mainly treated by applying approximate models. Such were the studies of ion-molecule reactions [5.19-22] and inelastic (nonreactive) atom-molecule scattering [5.23-26]. However, the introduction (in the late 1960's) of the Trajectory Surface Hopping Model (TSHM) on the one hand, and the rapid development of efficient quantum mechanical methods to treat inelastic and reactive collisions on the other, evoked extensive activity in the direction of more rigorous treatments. In the TSHM, BJERRE and NIKITIN [5.27] coupled the Landau-Zener formula with the ordinary classical trajectory method and studied the quenching of an excited Na by N_2. Four years later, TULLY and PRESTON [5.28] extended this method to reactive systems and applied it to reactive charge transfer processes between D_2 and H^+. The calculated cross reactions agreed with experiment [5.29,30]. A different approach was used by MILLER and GEORGE [5.31] who developed a semiclassical method coupling the Stückelberg method for single coordinate with the classical trajectory method to be performed on a complex plane. This method was later applied to various systems [5.32, 33]. The treatment of electronic nonadiabatic transitions in atom-molecule collisions started not too long ago and the field has hardly been reviewed. Still, some aspects were mentioned by MICHA in 1975 [5.34]. The TSHM and related topics have been discussed by TULLY [5.35] and a unified picture of the field was recently given by CHILD [5.36].

5.1 The General Approach

Following the first semiclassical treatments, attempts were made to apply quantum mechanical methods to reactive systems undergoing electronic transitions. To be able to treat this case rigorously, one should repeat the Born-Oppenheim (BO) treatment and ignore the approximation Born and Oppenheim introduced. To do that, the Hamiltonian H written in the following form should be considered:

$$H = T_n + H_e \quad . \tag{5.1}$$

T_n is the nuclear kinetic energy and H_e is the electronic part containing the electronic kinetic energy and the electronic potential energy depending parametrically on the nuclear coordinates. If $\psi(e,n)$ is the total wave function of the electrons and the nuclei ("e" stands for electronic coordinates and "n" for nuclear coordinates), it can be presented in terms of an electronic basis set $\zeta_i(e;n_0)$; $i = 1,2,\ldots$, i.e.,

$$\Psi(e,n) = \sum_i \zeta_i\,(e;n_0)\chi_i(n) \quad . \tag{5.2}$$

In this expression, the $\chi_i(n)$; $i = 1,2,\ldots$, are the nuclear wave functions and $\zeta_i(e;n_0)$ are solutions of the eigenvalue problem

$$H_e\zeta_i(e;n_0) = V_i(n_0)\zeta_i(e;n_0) \quad i = 1,2,\ldots \tag{5.3}$$

where $V_i(n_0)$; $i = 1,2,\ldots$, are the corresponding electronic eigenvalues and n_0 stands for a set of nuclear coordinates which may or may not be equal to n. Next, we distinguish between two representations, namely, the diabatic [5.5,37] and the adiabatic [5.8].

5.1.1 The Diabatic Representation

In the diabatic representation, n_0 is assumed to be constant; thus, $\psi(e,n)$ is presented in terms of an electronic basis set which is unaffected by changes in the nuclear coordinates.

Recalling that Ψ is the solution of the equation

$$H\Psi = E\Psi \quad , \tag{5.4}$$

by substituting (5.2) in (5.4) and applying (5.3) we obtain

$$T_n\underset{\sim}{\chi} + (\underset{\approx}{U} - \underset{\approx}{E})\underset{\sim}{\chi} = 0 \tag{5.5}$$

where $\underset{\sim}{\chi}$ stands for a vector column of the functions $(\chi_1,\chi_2,\ldots)$ and $\underset{\approx}{U}$ is a potential matrix with the elements

$$U_{ij} = \langle\zeta_i|v(e;n_0) - v(e;n_0) + V_j(n_0)|\zeta_j\rangle \quad . \tag{5.6}$$

Here, v(e;n) is the total electronic potential made up of Coulomb potential terms which depend parametrically on the nuclear coordinates n. It is important to note that the bra and ket notations are applied with respect to integration over electronic coordinates both here and hereafter.

For simplicity we next consider a collinear system. The Schrödinger equation which governs the motion of three particles in a line is given in the following form:

$$\left\{-\frac{\hbar^2}{2\mu_{BC}}\frac{\partial^2}{\partial r^2} - \frac{\hbar^2}{2\mu_{A,BC}}\frac{\partial^2}{\partial R^2} + [U(r,R) - E]\right\}\chi(r,R) = 0 \tag{5.7}$$

where r and R are the vibrational and translational coordinates, μ_{BC} and $\mu_{A,BC}$ are the reduced masses of the diatomic molecule BC and the atom A with respect to BC, respectively. Accordingly, the corresponding equation, in case electronic transitions are included, is

$$\left[-\frac{\hbar^2}{2\mu_{AB}}\frac{\partial^2}{\partial r^2} - \frac{\hbar^2}{2\mu_{A,BC}}\frac{\partial^2}{\partial R^2} + U_{ii}(r,R) - E\right]\chi_i(r,R) + \sum_{i\neq j} U_{ij}(r,R)\chi_j(R) = 0 \quad . \tag{5.8}$$

In principle, to solve this set of equations one could apply the ordinary close coupling methods developed for the single surface case [5.38,39]. However, it should be emphasized that this can be done only when the off-diagonal elements U_{ij} are small enough, i.e.,

$$|U_{ij}| \ll |U_i - U_j| \quad . \tag{5.9}$$

Further work is necessary when these conditions are not fulfilled, due to numerical instabilities. We shall elaborate on this point later.

5.1.2 The Adiabatic Representation

In the adiabatic representation one applies an electronic basis set which depends on the nuclear configuration. This amounts to replacing n_0 by n in (5.2,3). The advantage of the adiabatic representation is two-fold:

(i) The electronic basis now depends on the nuclear coordinates; thus, it contains an important part of the nuclear information as well. Therefore, the need for an extended basis set such as is usually required in diabatic representations is reduced. Of course, when infinite sets are used the information is identical in diabatic and adiabatic representations. It is very likely that the need for an adiabatic basis set is much stronger in case of reactive systems with at least two asymptotic regions.

(ii) Based on (5.6) one sees that, by replacing n_0 by n, a diagonal potential matrix can be obtained and thus the numerical instabilities which we mentioned above can be avoided.

In addition, *ab initio* treatments such as CI calculations yield adiabatic surfaces and therefore, for such cases, the adiabatic approach is obligatory. The disadvantage in applying the adiabatic basis set lies in the fact that here T_n, which is a differential nuclear operator, also acts on the electronic wave function. In this way, the nonadiabatic coupling terms which cannot easily be calculated are formed. The nonadiabatic terms are responsible for the transitions between adiabatic surfaces; in this sense they are similar to the off-diagonal terms of the potential matrix in the diabatic representation.

For the sake of clarity, let us reconsider the collinear case [5.40]. First, we introduce the following notation:

(i) ∇ will stand for the vectorial operator

$$\nabla = \left(\frac{\partial}{\partial r} ; \frac{\partial}{\partial R}\right) \quad ; \tag{5.10}$$

ii) the scalar product A·B will be defined as

$$A \cdot B = A_r B_r + A_R B_R \tag{5.11}$$

where A_r (B_r) and A_R (B_R) are the vibrational and translational components of A (B). Also, the representation in terms of mass scaled coordinates for T_n will be preferred. Thus, replacing r and R in (5.7) by $\lambda^{-1}r$ and λR [where $\lambda = (\mu_{A,BC}/\mu_{BC})^{\frac{1}{4}}$] leads to

$$T_n = -\frac{\hbar^2}{2\mu}\left(\frac{\partial^2}{\partial r^2} + \frac{\partial^2}{\partial R^2}\right) \tag{5.12}$$

where

$$\mu = (\mu_{A,BC}\ \mu_{BC})^{\frac{1}{2}} = \left(\frac{m_A m_B m_C}{m_A+m_B+m_C}\right)^{\frac{1}{2}} \quad . \tag{5.13}$$

Writing $\psi(e,n)$ as

$$\psi(e,n) = \sum_i \zeta_i(e,n)\chi_i(n) \tag{5.14}$$

where

$$[H_e - V_i(n)]\zeta_i(e,n) = 0 \quad ; \quad i = 1,2\ldots \tag{5.15}$$

substituting (5.14) in (5.4), applying (5.15), multiplying from the left by $\zeta_j(e;n)$ and integrating over electronic coordinates yields

$$<\zeta_j|T_n \sum\chi_i|\zeta_i> + [V_j(n) - E]\chi_j = 0 \quad . \tag{5.16}$$

Since

$$T_n = -\frac{\hbar^2}{2\mu}\nabla^2 \quad , \tag{5.17}$$

it can be shown that:

$$<\zeta_j|T_n \sum_i \chi_i|\zeta_i> = -\frac{\hbar^2}{2\mu}\left(\nabla^2\chi_j + 2\sum_j \tau_{ji}^{(1)}\cdot\nabla\chi_i + \sum \tau_{ji}^{(2)}\chi_i\right) \tag{5.18}$$

where

$$\tau_{ji}^{(1)} = <\zeta_j|\nabla\zeta_i> \quad ; \quad \tau_{ji}^{(2)} = <\zeta_j|\nabla^2\zeta_i> \quad . \tag{5.19}$$

Combining (5.16-19) leads to the final set of equations expressed in a matrix form:

$$\nabla^2\underset{\sim}{\chi} + 2\underset{\approx}{\tau}^{(1)}\cdot\nabla\underset{\sim}{\chi} + \underset{\approx}{\tau}^{(2)}\underset{\sim}{\chi} = \frac{2\mu}{\hbar^2}(\underset{\approx}{V} - E)\underset{\sim}{\chi} \tag{5.20}$$

where $\underset{\approx}{V}$ is a diagonal matrix which contains the adiabatic potential energy surfaces.

Now comparing the two sets of equations, that for the diabatic case (5.8) and that for the adiabatic case (5.20), the latter undoubtedly seems more attractive.

5.1.3 The Adiabatic-Diabatic Transformation [5.8,40-42]

With (5.20) available, one should be able to simply apply the close coupling method [i.e., form a whole matrix from *each element* $\underset{\approx}{V}$, $\underset{\approx}{\tau}^{(1)}$ and $\underset{\approx}{\tau}^{(2)}$] and solve the resulting set of equations by some method. At this stage, however, severe difficulties are likely to arise. First, the equations contain first-order derivatives (the usual equations for the single surface case contain only second-order derivatives) and no efficient method for integration of these equations is available at present. Secondly, some of the $\underset{\approx}{\tau}^{(1)}$ matrix elements are known to vary with r and R rather sharply and this is likely to cause numerical instabilities in any further treatment. Cases such as these were encountered in the $(H + H_2)^+$ system [5.43](Fig.5.1) and in the $(Ar + H_2)^+$ system [5.44] (Fig.5.2). Moreover, for this kind of system, TOP and BAER [5.43] showed that the off-diagonal element $\tau^{(1)}_{r_{12}}$ (the vibrational nonadiabatic coupling element which couples the ground and the first excited electronic states) behaves asymptotically as

$$\lim_{R \to \infty} \tau_{r_{12}}(R,r) = \frac{\pi}{2}\,\delta(r - r_s) \quad . \tag{5.21}$$

Here, $r = r_s$ is the value of r where for $R \to \infty$ the two adiabatic surfaces come infinitely close to each other and, what amounts to the same thing, where the two diabatic surfaces cross each other. In order to exploit the efficiency of the diabatic representation and yet avoid the above-mentioned deficiencies, SMITH [5.8] devised a procedure which transforms the adiabatic representation of the Schrödinger equation (and its corresponding wave function) into a diabatic representation. Since the two representations contain exactly the same information, the solutions and the transition probabilities are identical. This procedure, originally suggested for the atom-atom case (a single internal coordinate case), was then extended by BAER [5.40,41] to the atom-molecule case (first to the collinear case and then to the three-dimensional), and by TOP and BAER [5.42] to reactive systems (a similar) method was also recently successfully applied to vibrational inelastic collisions [5.45]). In order to demonstrate the transformation for a case of more than one (internal) coordinate, we return in more detail to the collinear atom-molecule case involving two coordinates, R and r. Reconsidering (5.20), making the transformation

$$\underset{\sim}{\chi} = \underset{\approx}{A}\underset{\sim}{\eta} \tag{5.22}$$

and performing the required algebra, one obtains

$$\underset{\approx}{A}\nabla^2\underset{\sim}{\eta} + 2(\nabla\underset{\approx}{A} + \underset{\approx}{\tau}^{(1)}\underset{\approx}{A})\cdot\nabla\underset{\sim}{\eta} + \left[(\underset{\approx}{\tau}^{(2)} + \nabla^2 + 2\underset{\approx}{\tau}^{(1)}\cdot\nabla)\underset{\approx}{A} - \frac{2\mu}{\hbar^2}(\underset{\approx}{V} - E)\underset{\approx}{A}\right]\underset{\sim}{\eta} = 0 \quad . \tag{5.23}$$

The careful selection of the matrix $\underset{\approx}{A}$ will ensure that the coefficient of $\nabla\underset{\sim}{\eta}$ vanishes. This implies that $\underset{\approx}{A}$ has to be a solution of the equation

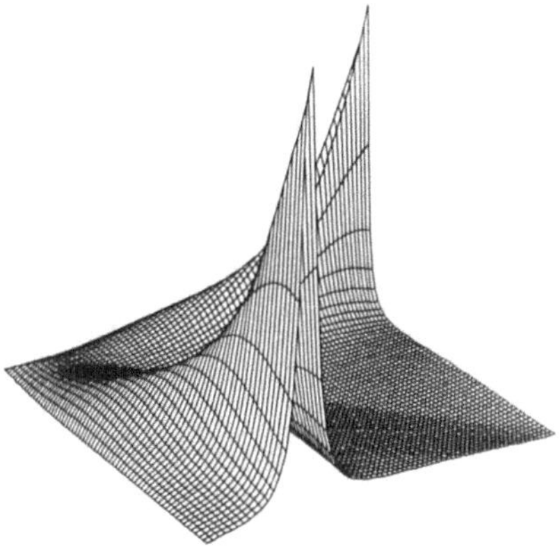

Fig.5.1. A three-dimensional figure for the *vibrational* nonadiabatic coupling term as a function of the interatomic distances in the $(H_2 + H)^+$ reactive system

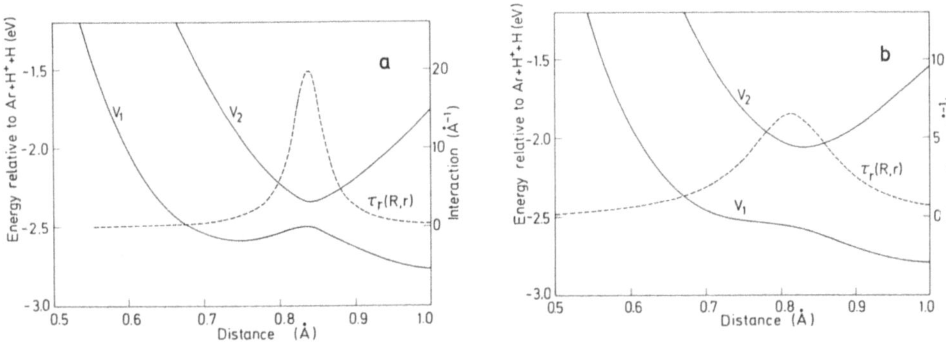

Fig.5.2a,b. The two lowest adiabatic energy levels $V_1(r,R)$ and $V_2(r,R)$ (——) and the corresponding vibrational nonadiabatic coupling term $\tau_r(R,r)$ (---) as a function of the interatomic distance $r = r_{HH}$ for a given fixed value of R. (a) R = 10 a.u.; (b) R = 8 a.u. The curves are for the $(Ar + H_2)^+$ system

$$\nabla \underset{\approx}{A} + \underset{\approx}{\vec{\tau}}^{(1)} \underset{\approx}{A} = 0 \quad . \tag{5.24}$$

This (vectorial) equation was shown to have a unique (and orthogonal) solution, provided the two components of $\underset{\approx}{\tau}^{(1)}$, i.e., $\underset{\approx}{\tau}_r^{(1)}$ and $\underset{\approx}{\tau}_R^{(1)}$, fulfilled the condition

$$\frac{\partial \underset{\approx}{\tau}_r^{(1)}}{\partial R} - \frac{\partial \underset{\approx}{\tau}_R^{(1)}}{\partial r} = \left[\underset{\approx}{\tau}_r^{(1)}, \underset{\approx}{\tau}_R^{(1)}\right] \quad . \tag{5.25}$$

Equation (5.24) can be replaced by an integral equation - a presentation which is convenient for computational purposes. One way of doing this (there are several other ways) is

$$\underset{\approx}{A}(r_1,R_1) = \underset{\approx}{A}(r_0,R_0) - \int_{R_0}^{R_1} \underset{\approx}{\tau}_R^{(1)}(r_0,R)\underset{\approx}{A}(r_0,R)dR - \int_{r_0}^{r_1} \underset{\approx}{\tau}_r^{(1)}(r,R_1)\underset{\approx}{A}(r,R_1)dr \quad , \tag{5.26}$$

where $\underset{\approx}{A}(r_0,R_0)$ can be chosen to be unity.

Returning to (5.23) and applying (5.24) it can be shown that the equation for $\underset{\sim}{\eta}$ simplifies dramatically:

$$\nabla^2 \underset{\sim}{\eta} - \frac{2\mu}{\hbar^2} (\underset{\approx}{W} - E)\underset{\sim}{\eta} = 0 \tag{5.27}$$

where $\underset{\approx}{W}$ is now a usual diabatic potential matrix given in the form

$$\underset{\approx}{W} = \underset{\approx}{A}^* \underset{\approx}{V}\, \underset{\approx}{A} \quad . \tag{5.28}$$

Thus, one may say that the adiabatic-diabatic transformation shifts the non-adiabatic coupling terms into the potential matrix. It should be emphasized that this diabatic potential matrix is different in general from the ordinary diabatic matrix which is based, as described above, on some (asymptotic) basis set. However, when the adiabatic basis set is large enough, the two coincide (this was demonstrated in a model study by BAER et al. [5.45]).

The two-state case [5.40] deserves a more detailed treatment because the equations simplify considerably. As mentioned, $\underset{\approx}{A}$ is orthogonal and can therefore be written as

$$\underset{\approx}{A}(r,R) = \begin{pmatrix} \cos\gamma & \sin\gamma \\ -\sin\gamma & \cos\gamma \end{pmatrix} \tag{5.29}$$

where $\gamma = \gamma(r,R)$. As the $\underset{\approx}{\tau}^{(1)}$'s are antisymmetric matrices, they take the form

$$\underset{\approx}{\tau}_R^{(1)} = \begin{pmatrix} 0 & \tau_R \\ -\tau_R & 0 \end{pmatrix} \quad ; \quad \underset{\approx}{\tau}_r^{(1)} = \begin{pmatrix} 0 & \tau_r \\ -\tau_r & 0 \end{pmatrix} \quad . \tag{5.30}$$

Substitution of (5.29,30) in (5.24) leads to

$$\frac{\partial\gamma}{\partial R} = -\tau_R \quad ; \quad \frac{\partial\gamma}{\partial r} = -\tau_r \tag{5.31}$$

and therefore $\gamma(r,R)$ becomes, see (5.26),

$$\gamma(r,R) = \gamma(r_0,R_0) - \int_{R_0}^{R} \tau_R(r_0,R)dR - \int_{r_0}^{r} \tau_r(r,R)dr \quad , \tag{5.32}$$

where $\gamma(r_0,R_0)$ can be assigned an arbitrary value.

Having constructed $\underset{\approx}{A}$, we can proceed by presenting the elements of $\underset{\approx}{W}$:

$$\begin{aligned} W_{11} &= V_1 \cos^2\gamma + V_2 \sin^2\gamma \\ W_{22} &= V_1 \sin^2\gamma + V_2 \cos^2\gamma \\ W_{12} &= W_{21} = (V_2 - V_1)\,\frac{1}{2}\sin 2\gamma \quad . \end{aligned} \tag{5.33}$$

Together with (5.32), (5.33) indicates that there are some difficulties with this method and that it should be modified. To emphasize the difficulties, let us consider the case where the nonadiabatic coupling terms are nonzero in a certain region in configuration space and zero elsewhere. Also, let us assume that the two adiabatic surfaces are reasonably close to each other in that region but sharply diverge upon leaving it so that the difference $V_2 - V_1$ becomes very large. Now, since the nonadiabatic coupling elements are nonzero in a certain region, γ is nonzero in that region and remains nonzero upon leaving it, see (5.32). From (5.33) it can be seen that W_{12} is nonzero in the above-mentioned region but starts to increase with the difference $(V_2 - V_1)$ upon leaving that region, thus increasing the coupling between the two diabatic surfaces. This strong coupling may cause severe instabilities in the numerical integration of the Schrödinger equation. In fact, it is entirely unnecessary because the system now moves in a "zero-interacting region" where the two adiabatic surfaces are uncoupled. In order to avoid situations such as that one or, more generally, in order to avoid the increase of the off-diagonal term in any case, an additional transformation [5.43,44] is required.

Considering (5.27), let us define a new nuclear set $\underset{\sim}{\xi}$, related to $\underset{\sim}{\eta}$ through a (constant) orthogonal transformation $\underset{\approx}{A_0}$, i.e.

$$\underset{\sim}{\eta} = \underset{\approx}{A_0} \underset{\sim}{\xi} \quad . \tag{5.34}$$

Substitution of (5.34) in (5.27) leads to

$$\nabla^2_{\underset{\sim}{\xi}} + \underset{\approx}{\bar{W}} \underset{\sim}{\xi} = 0 \tag{5.35}$$

where

$$\underset{\approx}{\bar{W}} = \underset{\approx}{A_0^*} \underset{\approx}{W} \underset{\approx}{A_0} \quad . \tag{5.36}$$

A_0 can now be chosen so that the off-diagonal elements of $\underset{\approx}{\bar{W}}$ diminish in value or disappear altogether. For instance, in the two-state case, $\underset{\approx}{A_0}$ takes the form

$$\underset{\approx}{A_0} = \begin{pmatrix} \cos \gamma_0 & -\sin \gamma_0 \\ \sin \gamma_0 & \cos \gamma_0 \end{pmatrix} \tag{5.37}$$

and the elements of $\underset{\approx}{\bar{W}}$ consequently become [5.43,44]

$$\bar{W}_{11} = V_1 \cos^2(\gamma - \gamma_0) + V_2 \sin^2(\gamma - \gamma_0)$$

$$\bar{W}_{22} = V_1 \sin^2(\gamma - \gamma_0) + V_2 \cos^2(\gamma - \gamma_0)$$

$$\bar{W}_{12} = \bar{W}_{21} = (V_2 - V_1) \frac{1}{2} \sin 2(\gamma - \gamma_0) \quad . \tag{5.38}$$

If γ happens to become a constant, say $\bar{\gamma}$, as described above, and if $(V_2 - V_1)$ starts to increase, the proper choice for γ_0 will be $\gamma_0 = \bar{\gamma}$. This will ensure the total disappearance of the off-diagonal and will retransform the diabatic into the adiabatic surfaces (but these surfaces will now be uncoupled).

As already stated, this procedure is a general method for dealing with large off-diagonal terms. It should be emphasized that this transformation should be employed before the close coupling expansion is applied. This is essential for the method to work.

As presented so far, the method seems to be relevant only when the electronic information is obtained from *ab initio* treatments. This is not the case and in fact its application cannot be avoided in many other cases where the potential is obtained by semiempirical methods (or by any other methods). As an example, let us consider the diatom-in-molecule (DIM) method [5.46-54]. Here one constructs a potential matrix with dimensions equal to the number of electronic states that are assumed to be relevant. The most direct procedure is applying the ordinary close coupling method and solving the corresponding set of differential equations. Such an approach may or may not yield results (depending on the dimensionality of the potential matrix) but in many cases it is not efficient. Alternatively, one could diagonalize the potential matrix (the eigenvalues are the adiabatic potential energy surfaces) and form the corresponding nonadiabatic coupling terms. Thus, the scattering problem would be treated within the framework of the adiabatic representation for which the dimensionality of the problem can be significantly reduced. To apply the adiabatic-diabatic transformation which will bring us back to a diabatic representation but with reduced dimensionality, the nonadiabatic coupling terms are needed. These can be obtained using the Hellmann-Feynman theorem in the form [5.48]

$$\tau_{x_{ij}} = \frac{\underset{\sim}{C}_i^* \frac{\partial \underset{\approx}{U}}{\partial X} \underset{\sim}{C}_j}{(V_i - V_j)} \quad ; \quad x = r,R \tag{5.39}$$

where $\underset{\sim}{C}_i$ and $\underset{\sim}{C}_j$ are the ith and jth eigenvectors of the DIM potential matrix and V_i and V_j are the corresponding eigenvalues. Once the adiabatic potential surfaces and the corresponding nonadiabatic coupling terms are available, the lowest adiabatic states (or any other states) can be selected and the relevant scattering problem solved.

5.1.4 The Adiabatic-Diabatic Transformation in Three Dimensions

So far, we have presented the adiabatic-diabatic transformation for the general case and applied it to the collinear (nonreactive) system. It is of interest and of importance to see how this transformation evolves in a more realistic situation and whether it still simplifies the equations.

The Hamiltonian for the nuclei and electrons is written as usual, i.e.,

$$H = T_n + T_e + v(e;n) \tag{5.40}$$

where T_e and $v(e;n)$ are given as before; T_n should, however, be considered in more detail. T_n is the kinetic energy operator which takes different forms according to the system of coordinates applied. Usually, two such systems, namely, the space-fixed and the body-fixed systems, can be distinguished. The two have their advantages and disadvantages and we shall not elaborate on those. However, when electronic transitions are included in the scattering process, the body-fixed system is far preferable because the electronic wave functions are expressed in terms of these coordinates. The starting point is the space-fixed system and its transformation to the body-fixed system. This transformation should, however, include the kinetic effect of the electron cloud which moves along with the atom-molecule system. The transformation has been done by BAER [5.41] and the corresponding form for T_n is

$$T_n = -\frac{\hbar^2}{2\mu}\left(\frac{1}{r^2}\frac{\partial}{\partial r}r^2\frac{\partial}{\partial r} + \frac{1}{R^2}\frac{\partial}{\partial R}R^2\frac{\partial}{\partial R}\right) + \frac{1}{2\mu}\left(\frac{1}{r^2}+\frac{1}{R^2}\right)j'^2 + \frac{1}{2\mu R^2}(\vec{K}^2 - \vec{K}z^2 + \vec{K}_+\vec{j}'_- + \vec{K}_-\vec{j}'_+) \quad . \tag{5.41}$$

Here $\vec{K}$ is the total angular momentum of the system, namely,

$$\vec{K} = \vec{j} + \vec{\ell} + \vec{L} \quad , \tag{5.42}$$

where $\vec{j}$ is the internal angular momentum, $\vec{\ell}$ is the orbital angular momentum and $\vec{L}$ is the total electronic angular momentum. One should be somewhat careful with respect to $\vec{j}$ which differs from $\vec{j}'$ by $\overrightarrow{L_z}$; thus,

$$\vec{j}' = \vec{j} + \overrightarrow{L_z} \quad . \tag{5.43}$$

This makes $\overrightarrow{j'_z}$ equal to $\overrightarrow{K_z}$, as in the case of absence of electrons. The other operators, $\vec{K}_+$, $\vec{j}'_+$, $\vec{K}_-$ and $\vec{j}'_-$ are the usual raising and lowering operators, respectively.

Our next step is to consider the Schrödinger equation:

$$(T_n + H_e)\psi^K_M = E\Psi^K_M \quad , \tag{5.44}$$

where ψ^K_M are eigenfunctions of $\vec{K}^2$ and $\vec{K}_z$. Expressing Ψ^K_M in terms of $D^K_{MM'}$, the coefficients of the irreducible representation of the rotation group introduces $\Psi^K_{M'}$ which is the wave function in terms of body-fixed coordinates (M and M' are the z components of K along the space-fixed and the body-fixed z axis, respectively):

$$\Psi^K_M(e,n) = \sum_{M'} D^K_{MM'}(\Lambda)\Psi^K_{M'}(e,n) \tag{5.45}$$

where Λ stands for the three Euler angles. Substituting (5.45) in (5.40), applying the relations

$$\vec{K}^2 D^K_{MM'} = \hbar^2 K(K+1) D^K_{MM'} \quad , \quad \vec{K}_z D^K_{MM'} = \hbar M' D^K_{MM'}$$
$$\vec{K}_\pm D^K_{MM'} = \hbar\lambda^{KM'}_\pm D^K_{MM'} \tag{5.46}$$

where

$$\lambda_{\pm}^{KM'} = \sqrt{K(K + 1) - M'(M' \pm 1)} \quad, \tag{5.47}$$

using the fact that the $D_{MM'}^{K}$ are linearly independent and replacing $\psi_{M'}^{K}(e,n)$ by

$$\psi_{M'}^{K}(e,n) = \frac{1}{Rr} \phi_{M'}^{K}(e,n) \quad, \tag{5.48}$$

we finally obtain [5.55,56]

$$T_{MM}\phi_{M}^{K} + T_{MM+1}\phi_{M+1}^{K} + T_{MM-1}\phi_{M-1}^{K} + \frac{2\mu}{\hbar^2} (H_e - E)\phi_{M}^{K} = 0 \tag{5.49}$$

where the prime sign is deleted from M. Here,

$$T_{MM} = -\left(\frac{\partial^2}{\partial R^2} + \frac{\partial^2}{\partial r^2}\right) + \frac{1}{R^2} [K(K + 1) - 2M^2]$$

$$- \left(\frac{1}{r^2} + \frac{1}{R^2}\right)\left(\frac{\partial^2}{\partial \theta^2} + \cot \theta \frac{\partial}{\partial \theta} - \frac{M^2}{\sin^2 \theta}\right)$$

$$T_{MM\pm1} = -\frac{1}{R^2} \lambda_{\pm}^{KM} \left[(M \pm 1) \cot \theta \pm \frac{\partial}{\partial \theta}\right] \tag{5.50}$$

and θ is defined as the angle between $\vec{R}$ and $\vec{r}$.

In order to attain our goal without unnecessary complications, we assume below that the total electronic angular momentum is zero and remains zero during the collisions. In this way, all angular momentum of the system is due to the nuclei and therefore $\vec{K}$ becomes equal to $\vec{J}$ ($= \vec{\ell} + \vec{j}$). Applying this assumption, we may expand $\phi_{M}^{K}(e;r,R,\theta)$ in terms of the electronic basis set:

$$\phi_{M}^{K}(e;r,R,\theta) = \sum_{j} \zeta_{j}(e;r,R,\theta)\chi_{j}^{KM}(r,R,\theta) \quad, \tag{5.51}$$

where the ζ_{j}'s are the eigensolutions of

$$\left(H_e - V_j(r,R,\theta)\right)\zeta_{j}(e;r,R,\theta) = 0 \qquad j = 1,2,... \quad . \tag{5.52}$$

Substituting (5.51) in (5.49,50), applying (5.52), multiplying from the left by electronic wave function and integrating over electronic coordinates leads to

$$T_{MM}\chi^{KM} + T_{MM+1}\chi^{KM+1} + T_{MM-1}\chi^{KM-1} + \frac{2\mu}{\hbar^2} (\underset{\approx}{V} - E)\chi^{KM}$$

$$+ 2\underset{\approx}{\tau}^{(1)} \cdot \nabla\chi^{KM} + \underset{\approx}{\tau}^{(2)}\chi^{KM} + \underset{\approx}{\tau}Q\chi^{KM} = 0 \quad, \tag{5.53}$$

where ∇ is a vectorial operator

$$\nabla = \left(\frac{\partial}{\partial R} , \frac{\partial}{\partial r} , \frac{\sqrt{R^2 + r^2}}{rR} \frac{\partial}{\partial \theta}\right) \quad . \tag{5.54}$$

Q is an operator defined as

$$Q\underset{\sim}{\chi}^{KM} = \frac{1}{R^2}\left(\lambda_+^{KM}\underset{\sim}{\chi}^{KM+1} - \lambda_-^{KM}\underset{\sim}{\chi}^{KM-1} + \frac{R^2 + r^2}{r^2}\cot\theta\underset{\sim}{\chi}^{KM}\right) \tag{5.55}$$

and the elements for $\underset{\approx}{\tau}^{(1)}$, $\underset{\approx}{\tau}^{(2)}$ and $\underset{\approx}{\Gamma}$ are:

$$\tau_{ij}^{(1)} = -\langle\zeta_i|\nabla\zeta_j\rangle \quad ; \quad \tau_{ij}^{(2)} = -\langle\zeta_i|\nabla^2\zeta_j\rangle \quad ;$$
$$\Gamma_{ij} = -\langle\zeta_i|\frac{\partial}{\partial\theta}\zeta_j\rangle \quad . \tag{5.56}$$

Here, as before, the scalar product A·B between two vectors is defined:

$$A\cdot B = A_R B_R + A_r B_r + A_\theta B_\theta \quad . \tag{5.57}$$

We now continue the procedure as in the collinear case; thus, an adiabatic-diabatic transformation $\underset{\approx}{A}$ matrix is introduced together with a new set of nuclear wave functions $\underset{\sim}{\eta}^{KM}$:

$$\underset{\sim}{\chi}^{KM} = \underset{\approx}{A}\underset{\sim}{\eta}^{KM} \quad . \tag{5.58}$$

Assuming that $\underset{\approx}{A}$ is a solution of the vector equation (and therefore is also orthogonal)

$$\nabla\underset{\approx}{A} - \underset{\approx}{\tau}^{(1)}\underset{\approx}{A} = 0 \quad , \tag{5.59}$$

it can be shown [5.41] that the corresponding equation for η^{KM} is

$$T_{MM}\underset{\sim}{\eta}^{KM} + T_{MM+1}\underset{\sim}{\eta}^{KM+1} + T_{MM-1}\underset{\sim}{\eta}^{KM-1} + \frac{2\mu}{\hbar^2}(\underset{\approx}{W} - E)\underset{\sim}{\eta}^{KM} = 0 \tag{5.60}$$

where $\underset{\approx}{W}$ is now the diabatic potential matrix obtained by

$$\underset{\approx}{W} = \underset{\approx}{A}^*\underset{\approx}{V}\,\underset{\approx}{A} \quad . \tag{5.61}$$

Thus, like in the collinear case, the nuclear kinetic operator used in single-surface problems remains unchanged when the number of electronic states is greater than one and the diabatic potential matrix $\underset{\approx}{W}$ is obtained, by means of a similar orthogonal transformation, from a diagonal adiabatic potential matrix. This outcome was obtained by ignoring altogether the total electronic angular momentum. Further work is necessary if a "nonzero electronic angular momentum case" is to be considered. In such a situation, the adiabatic electronic basis set and eigenvalues are insufficient to uniquely define the system. In addition, those functions would have to be expanded in terms of eigenfunctions of the electronic angular momentum. These eigenfunctions, in turn, are not always known, signifying that difficulties should be anticipated. For more details about this possibility, see the published work of DEVRIES and GEORGE [5.57] and MILLER and WYATT [5.58].

5.2 Numerical Examples

Collinear and three-dimensional treatments which include more than one electronic surface and are related to specific reactions are discussed in this section. We distinguish between reactive and nonreactive systems and as the author is better acquainted with the reactive systems, these will be presented in somewhat greater detail.

In this section we shall not refer to cases with one (mathematical) dimension, namely, potential curves (5.59). These will be discussed in Sect.5.3.

5.2.1 Reactive Systems

Several of the reactive treatments were carried out within the diabatic representation, justifying neither the origin of the diabatic surfaces, nor the origin of the diabatic coupling term. Therefore, in this sense all these studies can be considered at most as model studies. The quantum mechanical treatment of the electronic problem coupled with the existence of a reactive channel was first attempted by HAAS et al. [5.61] who formulated for this purpose a two-surface model similar to the one-surface model of HOFACKER and LEVINE [5.62]. They attributed electronic excitations to Franck-Condon factors which resulted from the different dynamic displacement of the reaction paths on the two electronic surfaces. The second study was by NAKAMURA [5.63] who performed a theoretical investigation of the conditions for electronic transitions using the distorted wave approximation formulation.

The first exact collinear reactive treatment on two surfaces was carried out by TOP and BAER [5.64,65]. In these model studies two H_3 type surfaces located one on top of the other and with one sometimes shifted with respect to the other, were coupled by a vibrational nonadiabatic coupling term which was either a constant or a Gaussian type function, peaking at the point where the two surfaces came the nearest to each other. As long as the adiabatic coupling was not too strong, the nuclear processes taking place on each surface could be decoupled from the electronic process. Consequently, the application of the distorted wave Born approximation (DWBA) was found most adequate. Transition probabilities obtained using the DWBA were found to be in excellent agreement with the exact ones. It was also verified that, as was assumed in early models, the vibronic states play a dominant role in the electronic transition from one surface to the other. For instance, the Massey parameter w [5.6], defined as

$$w = \frac{\Delta\varepsilon}{v\tau} \quad , \tag{5.62}$$

where $\Delta\varepsilon$ is the energy gap between two adiabatic curves in the atom-atom case, v is the radial velocity and τ, the adiabatic coupling term, is defined in the atom-molecule case between two *vibronic states* (one on each surface) and not between two *electronic surfaces*. Since the smaller w is, the larger is the expected transition probability, the Massey parameter tells us that if two vibronic states are nearing each other one may expect enhanced transition probabilities like in a regular resonance situation. The exact calculations for reactive systems [5.64,65] confirm this assumption.

Below, we give a detailed discussion of four reactive systems. The first two, $(H_2 + H)^+$ and $(Ar + H_2)^+$, were treated within the adiabatic framework and therefore were also exposed to the adiabatic-diabatic transformation discussed in the previous section. The other two, Ba + NO_2 and F + H_2, were treated within the diabatic framework.

a) The $(H_2 + H)^+$ System

To study the $(H_2 + H)^+$ system, TOP and BAER [5.43] used a DIM potential [5.48]. Figure 5.3 shows 3-dimensional drawings of the two lowest adiabatic surfaces and Fig.5.1 shows the *vibrational* nonadiabatic coupling term τ_r (the translational nonadiabatic coupling term is small and we ignore it for the rest of this discussion). It can be seen that the lower surface has a well of ~4 eV depth, whereas the upper surface has a barrier of 2 eV. The vibrational nonadiabatic coupling term is reasonably small for small R values but it increases sharply with R and, as was mentioned above, turns into a δ-function located at the vibrational coordinate value $r = r_s$ where the two diabatic surfaces intersect. This situation is shown in Fig.5.4 which gives the relative positions of the Morse potentials of the H_2 and H_2^+ molecules. Further details on the system are given in Fig.5.5a, where vibrational states of the two surfaces along the reaction coordinates are presented. Among other things, it is noted that the threshold for charge transfer is around 2 eV.

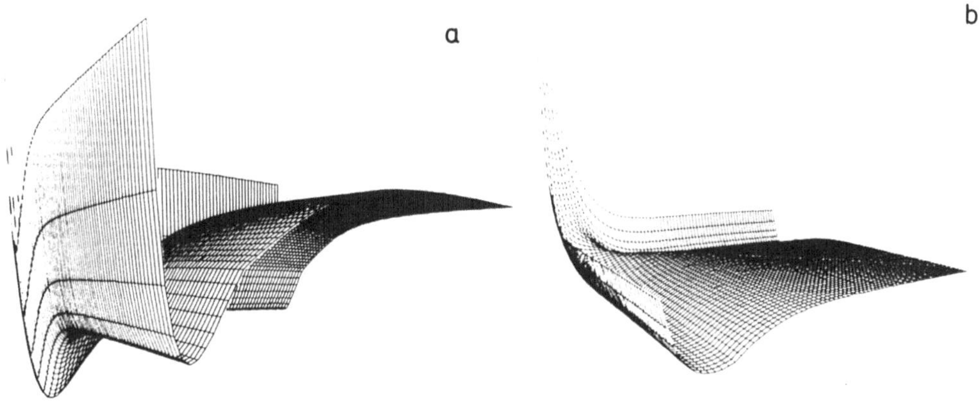

Fig.5.3a,b. Three-dimensional figures for the two lowest adiabatic surfaces for the reactive $(H_2 + H)^+$ system as a function of the interatomic distances. (a) The lowest surface; (b) the upper surface

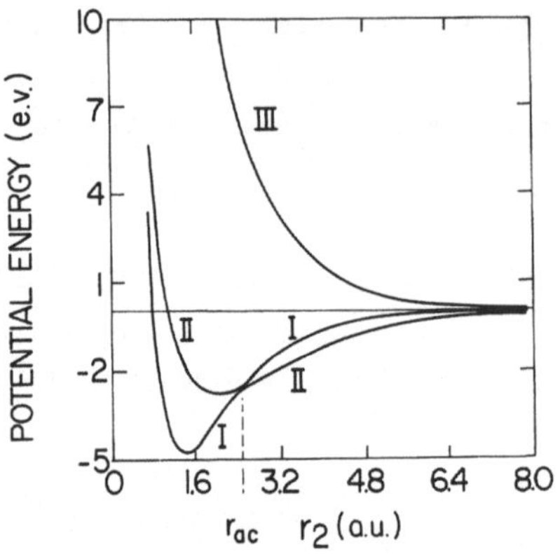

Fig.5.4. The three asymptotic diabatic (adiabatic) potential curves for the $(H_2 + H)^+$ system as a function of the diatomic distance. The value $r_2 = r_{ac}$ is the value where the seam is located and where γ (Fig.5.6) varies discontinuously from zero to $\pi/2$. The two lower curves present the diatomic potentials of H_2 and H_2^+

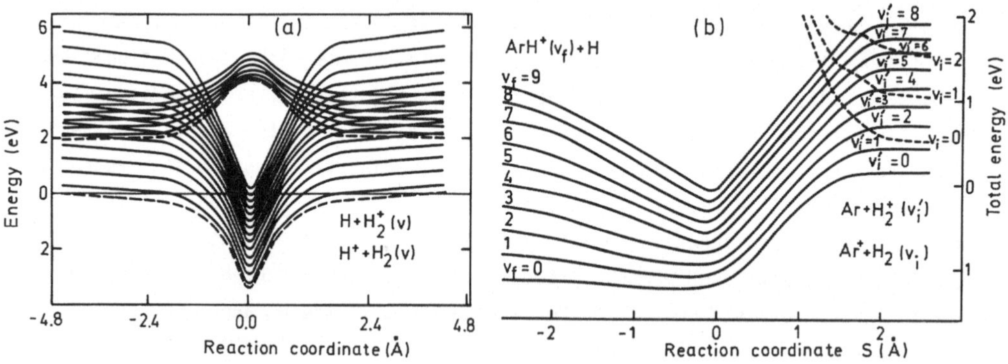

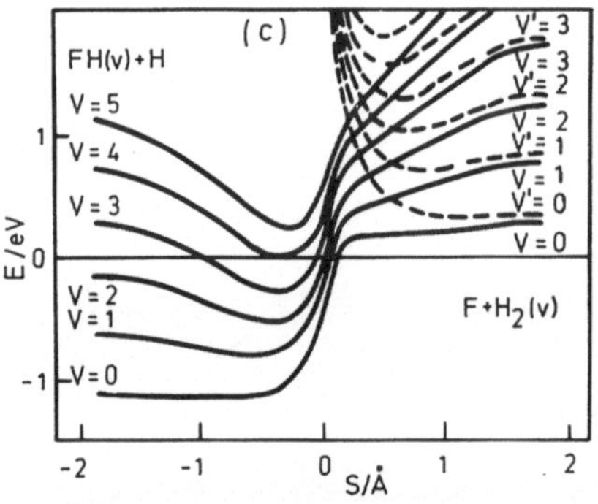

Fig.5.5a-c. The vibronic states as a function of the reaction coordinate. (a) The $(H + H_2)^+$ system; (b) the $(Ar + H_2)^+ \rightarrow ArH^+ + H$ system; (c) the $F(^2P_{3/2}, {}^2P_{1/2}) + H \rightarrow HF + H$ system

To obtain some further insight with respect to the rotation angle $\gamma(r,R)$, see (5.29), we present $\gamma(r,R)$ in Fig.5.6 as a function of r for different R values. As can be seen in the figure, the γ-function becomes more rectangular as R increases. This kind of behavior follows from the shape of $\tau_r(r,R)$ as R increases. Considering (5.32), ignoring the first integral since it is assumed that $\tau_R \sim 0$, and making $\gamma(r_0,R_0)$ equal to zero leads to

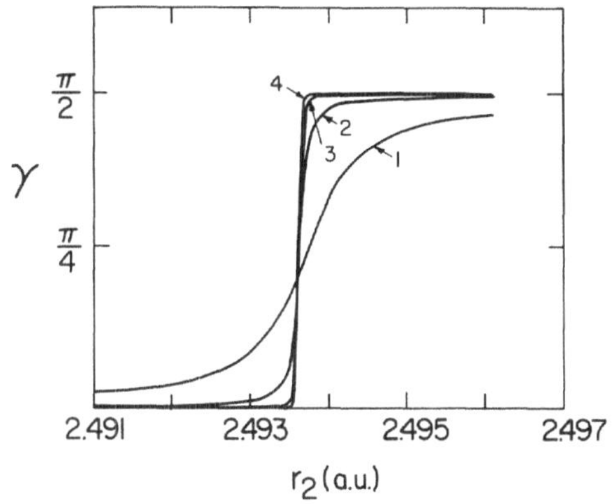

Fig.5.6. The adiabatic-diabatic transformation angle $\gamma(r,R)$ for the $(H_2 + H)^+$ system as a function of the vibrational coordinate r for various values of R. Curves 1-3 refer to three finite values of R (not large) and curve 4 presents the asymptotic case

$$\gamma(r,R) = \int_{r_0}^{r} \tau_r(r,R)dr \quad . \tag{5.63}$$

Substitution of τ_r given in (5.21) in (5.63) leads to a step function:

$$\lim_{R\to\infty} \gamma(r,R) = \frac{\pi}{2}\int_{r_0}^{r} \delta(r-r_s)dr = \begin{cases} 0 & r < r_s \\ \frac{\pi}{2} & r > r_s \end{cases} \tag{5.64}$$

which can clearly be seen in Fig.5.6.

Starting a reaction with a hydrogen molecule and a proton, four alternative reactions could result [provided that the energy exceed 2 eV, the minimal energy necessary to reach the upper surface (Fig.5.5a) yet still below dissociation]:

$$H_2(v_i) + H_I^+ \rightarrow \begin{cases} H_2(v_f) + H_I^+; & \text{vibrational inelastic} \\ H_2^+(v_f) + H_I; & \text{electronic nonadiabatic} \\ HH_1(v_f) + H^+; & \text{reactive} \\ HH_1^+(v_f) + H; & \text{reactive + electronic nonadiabatic} \quad . \end{cases} \tag{5.65}$$

TOP and BAER [5.43] presented results concerning all four channels. The results that will be given here for the purpose of illustration are concerned with electronic nonadiabatic processes and therefore no distinction will be made between reactive and nonreactive processes. Thus, $P_a(v_i,v_f)$ is the probability for transition from an initial state v_i to a final state v_f on the same surface ("a" stands for "adiabatic") and $P_{na}(v_i,v_f')$ is the probability for a transition to a final vibrational state v_f' on the upper surface ("na" stands for "nonadiabatic"). Having obtained these probabilities, we introduce the total adiabatic and nonadiabatic probability functions

$$P_a(v_i) = \sum_{v_f} P_a(v_i,v_f)$$

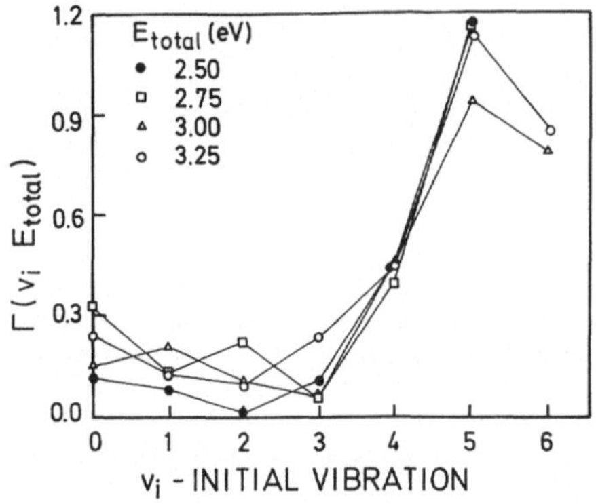

Fig.5.7. Electronic branching ratio for the $(H + H_2)^+$ system, $\Gamma(v_i) = P_{na}(v_i)/P_a(vi)$, as a function of the initial state v_i for different values of (total) energies

$$P_{na}(v_i) = \sum_{v'_f} P_{na}(v_i,v'_f) \tag{5.66}$$

and define the nonadiabatic branching ratio $\Gamma(v_i)$ as

$$\Gamma(v_i) = \frac{P_{na}(v_i)}{P_a(v_i)} \quad . \tag{5.67}$$

Figure 5.7 shows $\Gamma(v_i)$ as a function of v_i for different total energies. It is clear from the figure that $\Gamma(v_i)$ is much smaller than 1 for $v_i \leq 3$, but once $v_i \geq 4$, $\Gamma(v_i)$ is in the vicinity of 1. This result indicates the existence of vibrational-to-electronic resonance transitions for the higher vibrational states. By considering the vibrational states along the reaction coordinates (Fig.5.5a), $v_i \geq 4,5,6$ can be seen to be above the threshold for charge transfer and $v_i \leq 3$ below it.

b) The $(Ar + H_2)^+$ System

To study the $(Ar + H_2)^+$ system, BAER and BESWICK [5.44,66] used a DIM potential [5.67]. Four electronic states were included in the DIM matrix, whereas in the scattering calculation the two lowest adiabatic surfaces, together with the corresponding nonadiabatic coupling terms, were employed. The four states in the reagent asymptotic region are shown in Fig.5.8. Some information on coupling terms and surfaces is given in Figs.5.2 and 5.5b. Figure 5.2 shows the vibrational coupling term and the corresponding adiabatic potentials as a function of r for two different R values. It is seen that, as R increases, the function $\tau_r(R,r)$ peaks more sharply and the two adiabatic surfaces come closer to each other at the seam. Figure 5.5b shows the vibrational states of the two surfaces along the reaction coordinate. From this drawing the ArH^+ ion is seen to correlate with H_2^+ and so the reaction

$$Ar + H_2^+ \rightarrow ArH^+ + H \tag{5.68}$$

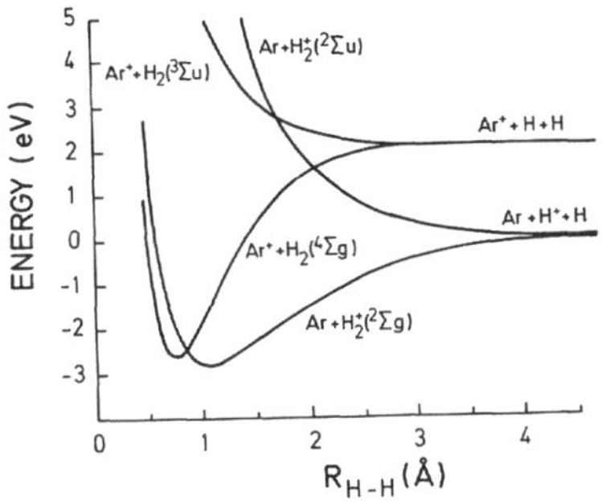

Fig.5.8

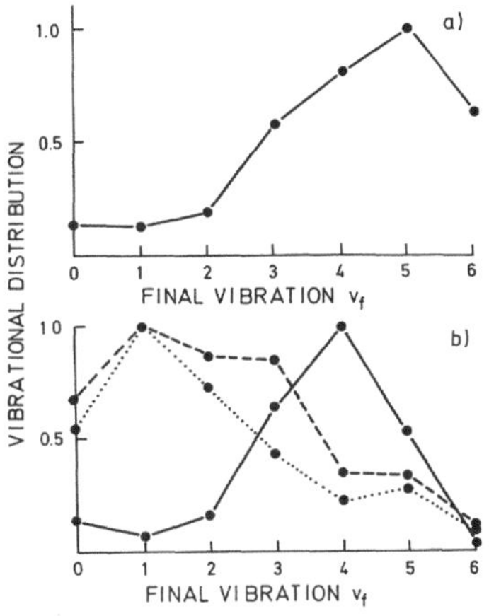

Fig.5.9a,b

Fig.5.8. The four asymptotic diabatic (adiabatic) potential curves for the $(Ar + H_2)^+$ system as a function of the diatomic distance. The two lowest curves present the diatomic potentials of H_2 and H_2^+

Fig.5.9. Normalized vibrational distribution of the ArH^+. (a) Results for the reaction $Ar^+ + H_2$ $(v_i = 0) \rightarrow ArH^+$ (v_f) + H which evolves through an electronic nonadiabatic transition (a two-surface calculation); (b) single-surface results for the adiabatic reaction Ar + H_2^+ $(v_i) \rightarrow ArH^+$ + H. $\cdots$ $v_i = 0$; --- $v_i = 1$; —— $v_i = 2$

is direct and does not involve an electronic nonadiabatic transition. On the other hand, the surface governing the reagents Ar^+ and H_2 is nonreactive and consequently, for the reaction

$$Ar^+ + H_2 \rightarrow ArH^+ + H \qquad (5.69)$$

to occur, an electronic nonadiabatic process should take place.

The two reactions are exothermic, the first by 1.30 eV and the second by 1.65 eV. The study by BAER and BESWICK emphasizes mainly the latter reaction. Threshold behavior and vibrational distribution were also treated.

The main findings were:

(i) the transition probability function demonstrates a threshold behavior at E_t = 0.06 eV;

(ii) a sharp spike in the probability function is encountered at E_t = 0.11 eV, indicating a possible complex with a lifetime of 10^{-12} s;

(iii) as for the vibrational distribution, the reaction is found to lead to a highly inverted population, peaking at v_f = 5 (Fig.5.9a).

Findings (i), (iii) are a result of the intersection of the $v_i = 0$ vibrational state of the $Ar^+ + H_2$ system with the $v_i' = 2$ of the $Ar^+ + H_2$ system. The threshold energy of 0.06 eV is due to a barrier of 0.06 eV formed as a result of the diabatic coupling between the two interacting states. This intersection will be discussed in further detail in Sect.5.4. The vibrational distribution is seen to be very similar to a vibrational distribution obtained starting with $v_i' = 2$ from the lower surface. The vibrational distributions for $v_i' = 0,1,2$ of the lower surface are also added for comparison and they support this outcome (Fig.5.9b).

Only the result concerning the long-lived complex seems to be unrelated to the electronic processes taking place in the entrance channel. It could be attributed, however, to the shallow potential wells formed by the higher vibrational states in the interaction region, which probably support bound states (Fig.5.5b). This possibility was recently supported by experimental evidence presented by BILOTTA et al. [5.68,69].

c) The Ba + N_2O System

The Ba + N_2O system was studied by BOWMAN et al. [5.70] following chemiluminescent type experiments performed by JONAH et al. [5.71] and WREN and MENZIGER [5.72]. In those experiments it was established that the excited $a^3\pi$ state of BaO is the dominant reaction product and the motivation for this study was to try to establish this result theoretically. Two competing reactions were considered:

$$\mathrm{Ba} + \mathrm{N_2O} \longrightarrow \begin{cases} \mathrm{BaO}(X^1\Sigma) + \mathrm{N_2} & \text{(singlet)} \\ \mathrm{BaO}(a^3\pi) + \mathrm{N_2} & \text{(triplet)} \end{cases} \quad . \tag{5.70}$$

To perform the study, the (Ba + N_2O) system was modelled in the following way:

(i) N_2 was considered as a single atom;

(ii) the experimental 4 eV exothermicity was replaced by 0.2 eV in order to avoid having to use a large number of vibrational basis functions;

(iii) the two surfaces v_s and v_t were chosen to be LEPS surfaces intersecting along the translational coordinate at a fixed value of $r = r_s$ (thus forming a seam), and the diabatic coupling potential was assumed to be different from zero in the vicinity of r_s, along R in the product channel. The second surface opened only in the product channel.

In addition to the quantum mechanical study, the authors [5.70] also performed calculations using the trajectory surface hopping method (TSHM) and were able to compare the results reached with the two methods.

Figure 5.10 shows reactive, electronically adiabatic and nonadiabatic transition probabilities. It can be seen that for the electronic adiabatic case the classical and the quantum mechanical results overlap reasonably well (at least as well as in single surface studies). However, for the nonadiabatic case, the two methods yield different results, raising questions as to the relevance of the TSHM in general.

d) The F + H_2 System

A study of the reactive F + H_2 system was carried out by ZIMMERMAN et al. [5.73]. It followed a study by ZIMMERMAN and GEORGE [5.74] on the inelastic version of the same system. The two surfaces in the reactive F + H_2 case are due to the spin-orbit

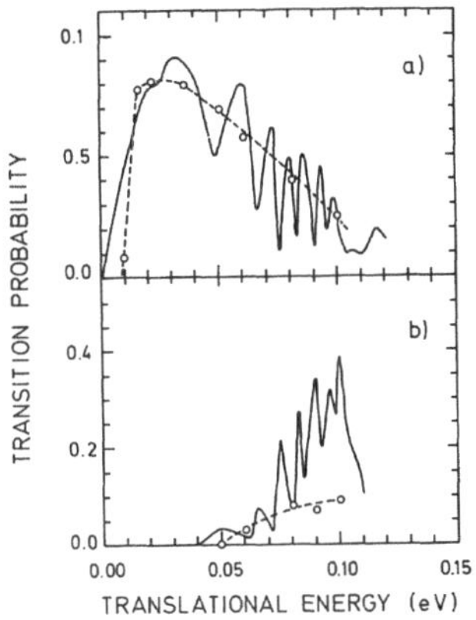

Fig.5.10a,b. Reactive transition probabilities for the system Ba + $N_2O \rightarrow BaO + NO$ as a function of translational energy. —— Quantum mechanical results; --- classical trajectory results. (a) Electronic adiabatic transition probability; (b) electronic nonadiabatic transition probability

coupling of the two ^2P levels of the fluorine atom, namely, $^2P_{3/2}$ and $^2P_{1/2}$. Consequently, the two reactions to be considered are

$$\left.\begin{array}{l} F(^2P_{3/2}) \\ \\ F(^2P_{1/2}) \end{array}\right\} + H \rightarrow HF + H \quad . \tag{5.71}$$

The ground electronic state V_1 was taken to be the Muckerman V potential (for the parameters of this potential see [5.75]). For the upper surface V_2, a modified version of the valence bond potential of BLAIS and TRUHLAR [5.76] was assumed. (This choice was done following a careful study of the three potential energy surfaces of this system [5.77].) The vibronic states of the two surfaces along the reaction coordinates are shown in Fig.5.5c.

The spin-orbit interaction was taken to be of the form

$$H_{so} = \frac{\lambda}{3}\begin{pmatrix} 1 & -\sqrt{2} \\ & \\ -\sqrt{2} & 2 \end{pmatrix} \quad , \tag{5.72}$$

where λ is the spin orbit splitting of the fluorine, i.e., $\lambda = 0.0502$ eV.

The published results [5.73] indicate that the upper surface sometimes affects the vibrational distribution starting the reaction with the triplet fluorine, and that the total reactive probability starting with the excited singlet fluorine is between 5 and 10%. However, the calculations were recently repeated because of some numerical instabilities which revealed themselves after the work was published. In fact, the effect of the upper surface was found to be very small (which means that the results with the triplet fluorine are almost identical to the single-surface results), and that the reactive probabilities with the $F(^2P_{1/2})$ are negligibly small.

Another study on the reactive F + H_2 system accompanied by electronic transition was also reported by WYATT and WALKER [5.78]. In a certain sense, their treatment is more general than the treatment described above since it applies to the three-dimensional case. However, since the authors considered only the entrance channel (they had used the same channel for their nonreactive study) and since it simulated the reactive channel by applying outgoing boundary conditions in the vicinity of the origin, theirs can be considered a model treatment. (It has limitations due to another reason, too: the authors assumed the reaction to be vibrational-adiabatic). Nevertheless, as far as the electronic nonadiabatic processes are concerned, their results are an upper limit for the effect of the upper surface since the electronic nonadiabatic coupling is located along the entrance channel only. Their findings seem to support the results of the collinear studies which we mentioned above, namely, negligibly small reactive cross sections for the reaction with the $F(^2P_{1/2})$. However, this conclusion could be somewhat misleading because the surfaces and the diabatic couplings applied in the two studies differed.

5.2.2 Nonreactive Systems

The first treatment of nonreactive systems was that by ZIMMERMAN and GEORGE [5.74] who considered the systems X + H_2; X = F, Cl, Br, I, and calculated the transition probability between two states which are asymptotically correlated with spin-orbit states of the halogen. The two states were coupled by the spin-orbit coupling, which was assumed to be constant as a function of the vibrational and translational coordinates. Such a coupling matrix was presented in (5.72) above.

Resonance conditions have been shown to be of major importance for electronic transitions. Consequently, the three systems X + H_2; X = F, Cl and I exhibited a high degree of vibrational elasticity. However, since the electronic spacing in the systems F + H_2 and Cl + H_2 is small, the two exhibited high activity in transition from one electronic state to the other. In case of iodine, the spacing is very high (~ 2 eV) and therefore the system exhibited electronic elasticity as well. The only exception was the H_2 + Br system which, due to the fact that its electronic and vibrational spacings are almost equal, was active vibrationally and electronically. Figure 5.11 compares the probabilities for transitions which are in resonance with those which are not.

The F + H_2 system was also treated in two dimensions by DeVRIES and GEORGE [5.79] and in three dimensions twice, i.e., by REBENTROST and LESTER [5.60] and by WYATT and WALKER [5.78]. In their study [5.60] REBENTROST and LESTER formulated their equations within an adiabatic framework and consequently calculated the SCF potential energy surfaces $1^2A'$, $2^2A'$ and A" and the corresponding electronic eigenfunctions [5.80]. The electronic basis set was then used to calculate the nonadiabatic coupling terms. Next, the adiabatic-diabatic transformation was applied, but only with regard to the lowest two surfaces and the rotation angle γ was not calculated as

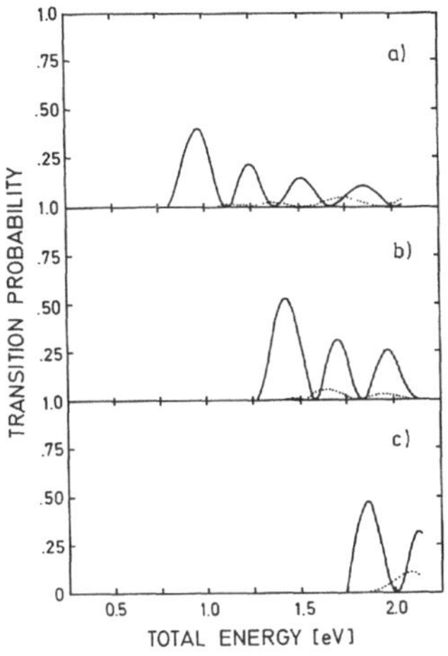

Fig.5.11a-c. Inelastic transition probabilities for the system $Br(^2P_{3/2}) + H_2(v_i) \rightarrow Br(^2P_{1/2}) + H_2(v_f)$ as a function of total energy. --- Transitions $v_i = m \rightarrow v_f = m$; —— transitions $v_i = m + 1 \rightarrow v_f = m$. (a) The case m = 0; (b) the case m = 1; (c) the case m = 2

rigorously as it should have been but assumed to be equal to Θ_F, the angle formed by $\vec{R}$ and the open shell orbital of the fluorine, without clear justification. The aim in this study was to determine the quenching cross section for the reaction [5.81]

$$H_2(j_i = 0) + F(^2P_{1/2}) \rightarrow H_2(j_f) + F(^2P_{3/2}) \quad . \tag{5.73}$$

Similar to the situation in the model calculations of TOP and BAER for reactive systems [5.64,65] and of ZIMMERMAN and GEORGE for nonreactive systems [5.74], it was found that the energy defect ΔE for the transition played a dominant role, namely, the smaller ΔE was, the larger the cross reaction became. Thus, in this sense at least, there was no difference between vibrational-to-electronic (VE) and rotational-to-electronic (RE) transitions. As far as numerical results were concerned, the cross section for reaction [5.73] was found to be between 10 and 16 $Å^2$ in the energy range of 0-2 kcal/mole; all other cross sections were at least one order of magnitude smaller.

The study by WYATT and WALKER [5.78] differed from the previous one, not only in the choice of a different potential surface but also in the formulation of the problem as a whole. From the start they presented their equations within the diabatic framework, using the DIM method to obtain all the required electronic information, namely, the diabatic potential matrix elements and the relevant electronic eigenfunctions. The advantage of this approach is that it enables a more straightforward construction of the relevant eigenfunction for the different electronic angular momentum operators. Its main drawback, however, is that during the scattering calculation, a large number of electronic states (24 in this case) have to be carried along. As in all previous treatments, here too the authors found the resonant transition probabilities to be the dominant ones. As for the corresponding cross sections, they found them to be one order of magnitude smaller than the Rebentrost-Lester cross section.

The excitation of C^+ ($^2P_{3/2}$) atoms as a result of collisions with H_2 (j = 0) was studied by CHU and DALGARNO [5.82]. Their treatment was much more restricted than the treatments described so far; they assumed the molecule H_2 as remaining in the j = 0 state during the collision and ignored various adiabatic and diabatic terms, thus oversimplifying their calculations. As for numerical results, they found the excitation cross sections to increase from ~1 to 25 A^2 in the energy range of 0.01 - 0.1 eV. The opposite process (quenching) was found to be half as efficient, on the average.

Electronic and rotational transition in the scattering of Na(^{2}P) by N_2 was studied by AMAEE and BOTCHER [5.83] who formulated their equation in the adiabatic representation and then transformed it into the diabatic one. For rotational excitation, they reported cross sections of about 100 A^2 in the energy range of 0.004 -0.05 eV.

Recently, MCGUIRE and BELLUM [5.84] studied the quenching process of Na(^{2}P) by hydrogen:

$$\mathrm{Na}(^2P) + H_2(v_i = 0) \rightarrow \mathrm{Na}(^2S) + H_2(v_f) \quad , \tag{5.74}$$

applying surfaces calculated by BOTSCHWINA and MAYER [5.85].

It was demonstrated that the product vibrational distribution is governed by a curve crossing between the vibrational ground state of the $^2B_2$ electronic state and the second vibrational state of the $^2A_1$ electronic state. Consequently, this state was found to be by far the most populated state, in accordance with experiment [5.86].

5.3 Electronic Nonadiabatic Processes Among Potential Curves

A network of potential curves usually stands for vibrational states related to one or more potential energy surfaces. The use of potential curves as a model for atom-molecule collision goes back to the early days of quantum mechanics. It started when PELZER and WIGNER [5.87] established the uniqueness and the importance of the minimum energy path in the H + H_2 potential surface generated by EYRING and POLANYI [5.88]. One year previously, ECKART [5.89] constructed a potential curve to study the penetration of electrons through a potential barrier and derived the corresponding transmission probability. The Eckart potential was flexible enough to fit also the minimum energy path of the H + H_2 surface and was taken over by WIGNER [5.90], BELL [5.91] and others for use in the study of chemical reactions. The idea of replacing atom-diatom surfaces by curves was considerably extended by BAUER, FISCHER and GILMORE (BFG) [5.23] who used it in studying the quenching reaction of Na(3^2P) by N_2. To explain the large experimental cross sections for this and other similar processes, BFG devised a network of potential curves related to three different potential energy surfaces: the initial, correlated with the excited Na, the final, correlated with the Na ground state Na (3^2S), and an intermediate ionic potential energy surface which was assumed to couple the two and correlated with Na^+ and N_2^-. The various curves form a grid of crossing points, each of which is attributed a diabatic and an adiabatic transition probability that was calculated using the Landau-Zener formula [5.2,3]

$$P_d = \exp(-q)$$

$$P_a = 1 - \exp(-q) \tag{5.75}$$

$$q = \frac{2\pi u^2}{\Delta F \hbar v} \quad . \tag{5.76}$$

Here, u is the diabatic coupling term, v is the velocity and ΔF stands for the difference in the gradients of the two intersecting curves:

$$\Delta F = |F_1 - F_2| \quad ; \quad F_i = \frac{\partial V_{ii}}{\partial R} \quad ; \quad i = 1,2 \quad , \tag{5.77}$$

all calculated at the crossing point. This model will be elaborated on in a section describing its application to reactive systems.

In Sect.5.3.1 we shall discuss the conditions for reducing a surface crossing problem to a curve crossing problem; in Sect.5.3.2 the two-curve crossing problem will be discussed to some extent, and in Sect.5.3.3 the theory will be extended to a network of curves.

5.3.1 Reducing the Surface Crossing Problem to a Curve Crossing Problem

Usually the curve crossing problem is treated within the diabatic framework and consequently we assume that the Schrödinger equation for the surface crossing problem is within the diabatic representation (for simplicity, the collinear case is considered):

$$-\frac{\hbar^2}{2\mu}\left(\frac{\partial^2}{\partial R^2} + \frac{\partial^2}{\partial r^2}\right)\chi_n + (W_{nn} - E)\chi_n + \sum_{n \neq n'} W_{nn'}\chi_{n'} = 0$$

$$n = 1,2,\ldots,N \quad . \tag{5.78}$$

The next step is expanding $\chi_n(R,r)$; $n = 1,2,\ldots$ in terms of a vibrational basis set. In dealing with a reactive system, the vibrational basis set should be adiabatic but when considering a nonreactive problem, this basis set may or may not be adiabatic. Since the adiabatic approach is somewhat more complicated, we continue with it, thus assuming the vibrational basis set to be parametrically R dependent:

$$\chi_n(r,R) = \sum_{i=1}^{M} \eta_{ni}(R)\phi_{ni}(r,R) \tag{5.79}$$

where $\phi_{ni}(r,R)$ are solutions of the eigenvalue problem

$$\left[-\frac{\hbar^2}{2\mu}\frac{d^2}{dr^2}+W_{nn}(r,R)-\varepsilon_{ni}(R)\right]\phi_{ni}(r,R)=0 \qquad \begin{array}{l} n=1,\dots,N \\ i=1,\dots,M \end{array} \tag{5.80}$$

(for simplicity M is assumed to be independent of n).

Substituting (5.79) in (5.78) and applying (5.80) leads to

$$-\frac{\hbar^2}{2\mu}\frac{d^2}{dR^2}\eta_n+(V_n-E)\eta_n+\sum_{n'}v_{nn'}\eta_{n'}-\left[2\tau_n^{(1)}\frac{d}{dR}+\tau_n^{(2)}\right]\eta_n=0 \quad , \tag{5.81}$$

where η_n is a column vector of the form

$$\eta_n=\begin{pmatrix}\eta_{n1}\\ \vdots \\ \eta_{nM}\end{pmatrix} \quad , \tag{5.82}$$

V_n is a diagonal matrix having the eigenvalues ε_{ni} in its diagonal, $v_{nn'}$ is a matrix with the elements of the form

$$v_{nin'i'}=\langle\phi_{ni}|W_{nn'}|\phi_{n'i'}\rangle \quad , \tag{5.83}$$

and $\tau_n^{(1)}$ and $\tau_n^{(2)}$ are the usual nonadiabatic terms:

$$\tau^{(1)}_{nii'}=\frac{\hbar^2}{2\mu}\langle\phi_{ni}\left|\frac{d}{dR}\phi_{ni'}\right.\rangle$$

$$\tau^{(2)}_{nii'}=\frac{\hbar^2}{2\mu}\langle\phi_{ni}\left|\frac{d^2}{dR^2}\phi_{ni'}\right.\rangle \quad . \tag{5.84}$$

Continuing now as in Sect.5.1.3, namely, replacing η_n by ζ_n,

$$\eta_n=A_n\zeta_n \tag{5.85}$$

where A_n is assumed to be a solution of

$$\frac{\hbar^2}{2\mu}\frac{d}{dR}A_n+\tau_n^{(1)}A_n=0 \quad , \tag{5.86}$$

one obtains

$$-\frac{\hbar^2}{2\mu}\frac{d^2}{dR}\zeta_n(R)+(\bar{W}_n-E)\zeta_n(R)+\sum_{n\neq n'}\bar{W}_{nn'}\zeta_{n'}=0 \tag{5.87}$$

where

$$\bar{W}_n=A_n^*V_nA_n \quad ; \quad \bar{W}_{nn'}=A_n^*V_{nn'}A_{n'} \quad . \tag{5.88}$$

For more details regarding this approach see [5.45].

Equation (5.87) is solved subject to the following boundary conditions:

$$\lim_{R \to \infty} \zeta_{ni}(R) = \delta_{nn_0}\delta_{ii_0}\, e^{-ik_{ni}R} + \sqrt{\frac{k_{n_0i_0}}{k_{ni}}}\, T^{(R)}_{nin_0i_0}\, e^{ik_{ni}R} \quad . \tag{5.89}$$

In dealing with a nonreactive problem, we have on the "other side":

$$\lim_{R \to 0} \zeta_{ni}(R) = 0 \tag{5.90}$$

but in a reactive problem we then have

$$\lim_{R \to -\infty} \zeta_{ni}(R) = \sqrt{\frac{k_{n_0i_0}}{k'_{ni}}}\, T^{(T)}_{nin_0i_0}\, e^{-ik'_{ni}R} \quad . \tag{5.91}$$

Here

$$\begin{aligned} k_{ni} &= \sqrt{\frac{2\mu}{\hbar^2}\,[E - \varepsilon_{ni}\;(R = \infty)]} \\ k'_{ni} &= \sqrt{\frac{2\mu}{\hbar^2}\,[E - \varepsilon_{ni}\;(R = -\infty)]} \quad ; \end{aligned} \tag{5.92}$$

the indices (R) and (T) on the T-matrix elements stand for "reflected" and "transmitted," respectively. Equation (5.87) can also be written in the form

$$-\frac{\hbar^2}{2\mu}\frac{d^2}{dR^2}\,\underset{\sim}{\zeta} + (\underset{\approx}{\bar{W}} - E)\underset{\sim}{\zeta} = 0 \quad , \tag{5.93}$$

where $\underset{\approx}{\bar{W}}$ stands for a "super" potential matrix and $\underset{\sim}{\zeta}$ for a "super" wave function vector.

As an example, we shall apply this approach to the $(Ar + H_2)^+$ system [5.44]. Since the electronic interaction takes place in the entrance channel only before the strong vibrational interaction region is reached (Fig.5.5b), we avoid using the above-mentioned adiabatic transformation and use only the asymptotic basis set (i.e., a diabatic basis set). The various diabatic potential matrix elements are:

$$\begin{aligned} \bar{W}_{11} &= 0.5 + 0.19\,\exp[-4.5(R - 4.4)] \\ \bar{W}_{22} &= 0.68\,\tanh[2.11(R - 3.5)] \\ \bar{W}_{33} &= \bar{W}_{22} + 0.24 \\ \bar{W}_{12} &= \bar{W}_{13} = 14.52\,\exp(-1.15R) \\ \bar{W}_{23} &= 0 \quad . \end{aligned} \tag{5.94}$$

The energies are given in eV and the distances in A. Here, $\bar{W}_{11}$ stands for the vibrational state $v_i = 0$ of the upper surface and $\bar{W}_{22}$ and $\bar{W}_{33}$ for $v_i = 2$ and $v_i = 3$

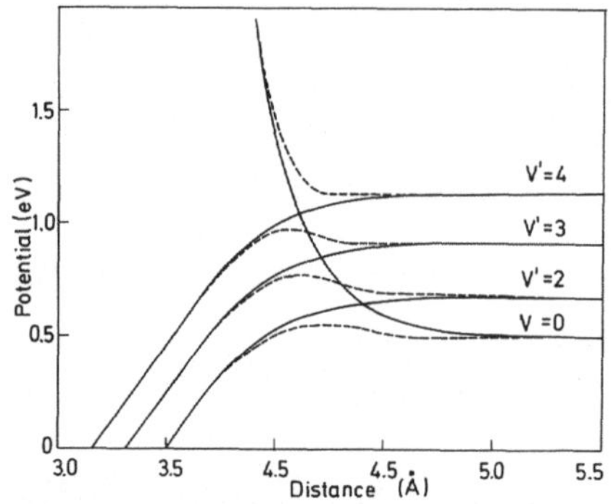

Fig.5.12

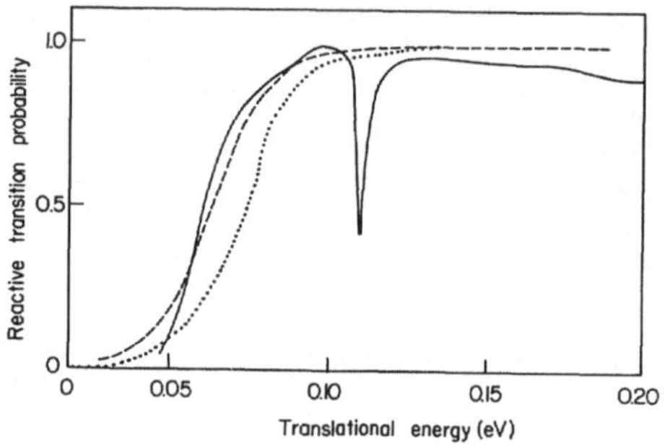

Fig.5.13

Fig.5.12. Diabatic and adiabatic vibronic states for the model studies of the $(Ar + H_2)^+ \rightarrow ArH^+ + H$ system as a function of translational energy

Fig.5.13. Reactive transition probabilities for the $(Ar + H_2)^+ \rightarrow ArH^+ + H$ system as a function of translational energy. —— The exact close coupling results; ··· results of a two-state model; --- results of a three-state model

of the lower surface, respectively (Fig.5.12). $\bar{W}_{33}$, which is the coupling between two states of the same surface ($v_i' = 2$ and $v_i' = 3$), is negligibly small in the asymptotic region and is therefore set equal to zero.

The set of differential equations is solved twice; once including only two curves (one for each surface), and once including three curves [one for the upper surface ($v_i = 0$) and two for the lower ($v_i' = 2,3$)]. Figure 5.13 presents the reactive probability functions for the process

$$Ar^+ + H_2(v_i = 0) \rightarrow ArH' + H \quad . \tag{5.95}$$

Three curves are shown, one obtained by the extensive close coupling calculations, i.e., the "exact", and the other two obtained by applying the two and three-curve model, respectively. For all practical purposes, the three-curve model results seem to overlap with the exact results.

5.3.2 The Two-Curve Model

The two-curve model was treated back in the early 1930's and these studies yielded very useful formulas which were general enough to be applied in most cases of interest. Such formulas can be found in the works by LANDAU [5.2] and ZENER [5.3] for the adiabatic transition probability, see (5.75), and a more general formula was given by STÜCKELBERG [5.4]. These treatments were further considered by BATES [5.13], NIKITIN [5.6] and CHILD [5.9,92,93]. A somewhat different situation was considered by DEMKOV [5.10]. Instead of viewing two diabatic curves crossing at a given point, DEMKOV considered two parallel curves with an interaction which increased sharply as the translation coordinate decreased from infinity. This case was further treated by NIKITIN [5.94] and by CROTHERS [5.95]. More details about the various treatments and how they related to each other are given in [5.36]. In this section we present a reactive two-curve model and discuss exact and approximate results [5.96].

The model was defined through two diabatic curves $V_1(R)$ and $V_2(R)$ which are coupled by a constant interaction term

$$V_1(R) = A\exp(-\alpha R) \quad ; \quad V_2(R) = A\exp(\alpha R) \quad . \tag{5.96}$$

Here, A is taken to be 0.1 eV and α to be 8 $Å^{-1}$, and the two functions were defined in the range $-\infty \leq R \leq \infty$. The equations to be solved were

$$\begin{aligned} &\left[-\frac{\hbar^2}{2\mu}\frac{d^2}{dR^2} + V_1(R) - E\right]\psi_1 + V_{12}\psi_2 = 0 \\ &\left[-\frac{\hbar^2}{2\mu}\frac{d^2}{dR^2} + V_2(R) - E\right]\psi_2 + V_{12}\psi_1 = 0 \quad . \end{aligned} \tag{5.97}$$

The calculations were performed as a function of μ, the mass of the system, and as a function of V_{12} which is the diabatic coupling term. The equations were solved exactly using a propagation method, and the reactive transition probabilities obtained in this way were compared with results obtained by various models - three of which were different versions of the distorted wave Born approximation (DWBA) and one a modified Landau-Zener formula [5.6].

The DWBA S matrix element S_{12} is given by

$$S_{12} = \langle\psi_1|V_{12}|\psi_2\rangle \tag{5.98}$$

where

$$-\frac{\hbar^2}{2\mu}\frac{d^2}{dR^2}\psi_i(R) + (V_i - E)\psi_i(R) = 0 \qquad i = 1,2 \quad . \tag{5.99}$$

In order to obtain an analytical expression for S_{12}, the potentials $V_i(R)$; $i = 1,2$ were approximated by a skew line potential for which the Airy function is the relevant solution. Applying this approximation, S_{12} can be obtained analytically [5.97]:

$$S_{12} = 2\pi V_{12}\left(\frac{\mu}{F^2\hbar^2}\right)^{1/3} A_i\left[-\Delta R\left(\frac{\mu F}{\hbar^2}\right)^{1/3}\right] \quad , \tag{5.100}$$

where F stands for the tangent of the two curves at their turning points:

$$F = E\alpha \quad (F_1 = -F_2 = F) \quad . \tag{5.101}$$

R is the distance between the two turning points:

$$\Delta R = R_2 - R_1 = \frac{2}{\alpha}\ln\left(\frac{E}{A}\right) \tag{5.102}$$

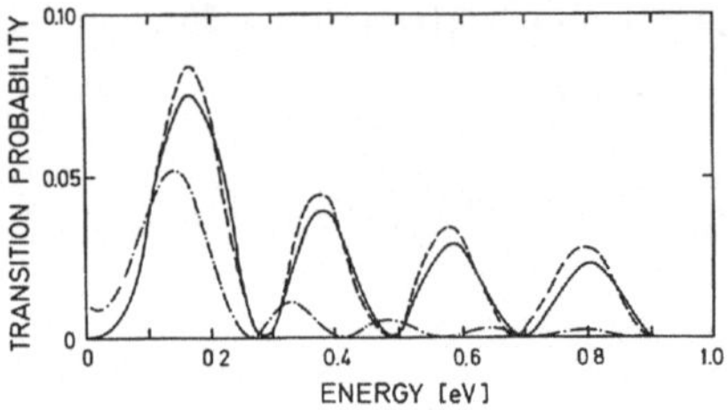

Fig.5.14. Reactive transition probabilities for the two-state model as a function of energy for m = 1.905 mu and V_{12} = 0.01 eV. —— Exact results; -·-·- DWBA results [(5.100); --- uniform DWBA results (5.105)]

and $A_i(x)$ is the Airy function. Thus, it can be seen that S_{12} becomes zero (resonance) at the zero of the Airy functions.

A comparison between this and the exact solution is shown in Fig.5.14 and, as can be seen, the agreement is poor. A considerable improvement is achieved when one takes a solution within the uniform approximation instead of an Airy function [5.98]:

$$\psi_j(R) = \sqrt{2}\pi z_j(R)^{1/4} k_j(R)^{-1/2} A_i[-z_j(R)] \; ; \quad j = 1,2 \tag{5.103}$$

where

$$k_j(R) = \sqrt{\frac{2\mu}{\hbar^2}[E - V_j(R)]}$$

$$z_j(R) = \left[\frac{3}{2}\int_{R_j}^{R} k_j(R)dR\right]^{2/3} \quad . \tag{5.104}$$

Applying this form of $\psi_j(R)$, the S_{12} element can still be obtained analytically [5.99]:

$$S_{12} = 4\pi\mu \frac{V_{12}}{\hbar^2}[2k_1k_2(k_1 - k_2)]^{-1/2}\rho^{1/4}A_i(-\rho) \tag{5.105}$$

where

$$\rho = \frac{3}{2}\left[\int_{R_1}^{0} k_1(R)dR + \int_{0}^{R_2} k_2(R)dR\right]^{2/3} \quad . \tag{5.106}$$

Again, Fig.5.14 compares the exact results with those obtained from this formula and; as can be seen, the fit is rather good. For much larger transition probabilities, the approximate (analytical) results deviate significantly but can be corrected to a certain extent by exponentiating the S matrix [5.100,101]. This amounts to replacing S_{12} with sin S_{12}.

Similar studies (on reactive models) were performed by LAING et al. [5.102] who considered a two-curve state model with the aim of studying tunneling and the effect of a closed state on the reactive transition probability. KORSCH and KRÜGER [5.103] derived a first-order approximation to the coupled equations in momentum representation and compared their results to the exact ones by BAER and CHILD [5.96] and by LAING et al. [5.102]. Good agreement was obtained in the two cases.

5.3.3 A Multi-Curve Crossing Model for Reactions

The treatment of a network of reactive curves which stand for the vibronic states of two surfaces crossing each other along a seam is presented in this section [5.104]. The purpose of this study is to test the reliability and determine the range of validity of two simplified models: the Bauer-Fischer-Gilmore (BFG) model [5.23] and a Franck-Condon type model.

a) The Model

The model, which contains two surfaces, is defined in terms of two displaced harmonic oscillators with equal force constants. As for the R dependence, one is assumed to be constant and the other to increase exponentially with R.

Thus,

$$\begin{aligned} V_1(r,R) &= \frac{1}{2}k(r-r_1)^2 \\ V_2(r,R) &= \frac{1}{2}k(r-r_2)^2 + A[1-\exp(R/R_0)] \quad . \end{aligned} \qquad -\infty \le R \le \infty \qquad (5.107)$$

The coupling is taken to be constant over the effective crossing region. Scaling the various energy and length variables and following the procedure described in Sect.5.3.1, one ends up with the following set of equations:

$$\left(\frac{1}{\rho_0^2}\frac{d^2}{dR^2} + E\right)\phi_{\nu i} = \sum_{\nu'\ell'} V_{\nu i\nu' i'}\phi_{\nu' i'} \quad , \qquad (5.108)$$

where E is the total energy (in units of $\hbar\omega$), ρ_0 is given by

$$\rho_0 = R_0/r_0 \quad , \qquad (5.109)$$

r_0 being half the width of the oscillator, ν and i being indices for the surfaces and the vibronic states, respectively, and $V_{\nu i\nu' i'}$ given as

$$V_{\nu i\nu' i'} = \begin{cases} i\delta_{ii'} & \nu = \nu' = 1 \\ [\alpha(1-e^{-\bar{\rho}}) + i]\delta_{ii'} & \nu = \nu' = 2 \\ U_0 S_{ii'}(\xi_0) & \nu \neq \nu' \end{cases} \quad . \qquad (5.110)$$

Here, U_0 is the scaled coupling term V_0, $S_{ii'}(\xi_0)$ is the overlap integral between two harmonic wave functions shifted by a scaled distance ξ_0 and α is equal to $A/\hbar\omega$.

Equations (5.108) were solved exactly by a propagator method subject to the usual (reactive) boundary conditions.

b) The Simplified Models

The Bauer-Fischer-Gilmore Model

The BFG model is a simple and straightforward extension of the Landau-Zener model to a network of curves. Thus, for each grid point (i,j) (the crossing of curve i belonging to surface 1, with curve j belonging to surface 2), one attributes a diabatic probability $P_d(i,j)$ and an adiabatic probability $P_a(i,j)$, both as given by the Landau-Zener formula, see (5.75). To continue, it is assumed that the system evolves from right ($R = \infty$) to left ($R = -\infty$) without reflection. If $\sigma_1(i,j)$ is the probability for the system to pass the grid point (i,j) and to be on surface 1 and $\sigma_2(i,j)$ is defined similarly but with respect to surface 2, then the equation for $\sigma_\nu(i,j)$, $\nu = 1,2$ (taking into account the previous corresponding probabilities) is

$$\begin{aligned}\sigma_1(i,j) &= \sigma_1(i,j-1)P_d(i,j) + \sigma_2(i+1,j)P_a(i,j)\\ \sigma_2(i,j) &= \sigma_1(i,j-1)P_a(i,j) + \sigma_2(i+1,j)P_d(i,j)\quad .\end{aligned} \tag{5.111}$$

The chain of equations just presented is solved subject to a given set of initial (boundary) conditions.

The Franck-Condon (FC) Model

In the context of the Franck-Condon model, one assumes that the transition from one surface to the other is sudden, with no intermediate transitions as in the BFG model, and that the system moves from a given vibrational state in one surface to some vibrational state of the other. The transition probability, therefore, involves a Landau-Zener branching ratio between the *surfaces*, multiplied by an appropriate Franck-Condon overlap factor:

$$P^a_{I,j} = P_a|S_{I,j}|^2 \tag{5.112}$$

where I stands for the initial vibronic state on surface 1. As mentioned, P_a is a Landau-Zener probability function, calculated at the point where the two curves I and j intersect.

In general, the two models yield different results (their results become identical when either the coupling term or the overlap integral becomes small enough) and the question is whether a single parameter can be found such as to enable determination of the range of validity of each model. Obviously such a parameter is the energy. The FC model becomes more relevant at higher energies. It is expected that the BFG model is more relevant for low energies because it is based on a step-by-step transition process. However, this parameter is not sharp enough because the relevance of the models is also related to the way the energy is distributed among the various degrees of freedom as well as on the relative position of the seam at the crossing point. CHILD and BAER [5.104] were able to construct a parameter γ in which all this information was lumped together.

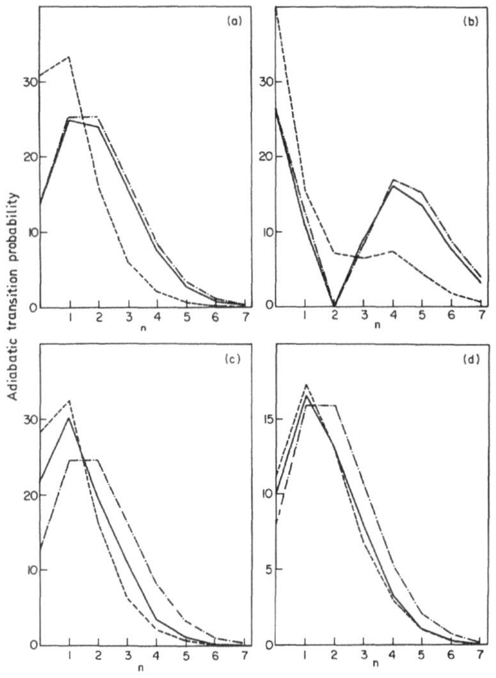

Fig.5.15a-d. Reactive transition probabilities for the multicurve crossing model. —— Exact results; --- Bauer-Fischer-Gilmore model results; -·-·- Franck-Condon model results. (All the calculations were done for the same total energy.) (a) The probability for the (reactive) transition ($\nu = 1$, $i = 0$) $\to$ ($\nu' = 2$, $i' = n$). The value of γ is 50; the total nonadiabatic transition probability is 0.9. (b) The probability for the (reactive) transition ($\nu = 1$, $i = 1$) $\to$ ($\nu' = 2$, $i' = n$). The value of γ is 50; the total nonadiabatic transition probability is 0.9. (c) The probability for the (reactive) transition ($\nu = 1$, $i = 0$) $\to$ ($\nu' = 2$, $i' = n$). The value of γ is 2; the total nonadiabatic probability is 0.9. (d) The probability for the reactive transition ($\nu = 1$, $i = 0$) $\to$ ($\nu' = 2$, $i' = n$). The value of γ is 2; the total nonadiabatic probability is 0.5

Considering the equation of the seam

$$G(r,R) = V_1(r,R) - V_2(r,R) = 0 \quad , \tag{5.113}$$

it was shown that γ takes the form

$$\gamma = \frac{v_R}{v_r}\frac{\Delta G_R}{\Delta G_r} \tag{5.114}$$

where

$$\Delta G_s = \frac{\partial G}{\partial s} \quad ; \quad s = r,R \tag{5.115}$$

and v_R and v_r are the translational and vibrational velocities, respectively. Next, it was established that when $\gamma \gg 1$ the FC model becomes exact, but when $\gamma \ll 1$ the BFG model should be preferred.

Figure 5.15 compares the exact results with those obtained with the models and the parameter γ can be seen to enable the relevance of the two models to be distinguished. (All the calculations were done for the same energy.)

5.4 Electronic Nonadiabatic Processes in Strong Laser Fields

The study of electronic nonadiabatic transitions caused by strong radiation fields (laser radiation) operating during a molecular collision was recently introduced by ZIMMERMAN et al. [5.105]. Following the stimulating research on the effect of laser fields on atomic collisions [5.106-108], the above authors performed their study on atom-molecule collisions with a particular emphasis on those instances in which vib-rotational level spacing within the collision complex passes through a resonance with the field quantum, as the encounter proceeds. Although no experimental evidence has been published regarding the effect of a laser field on an atom-diatom collision process, still it is of interest to briefly discuss here the main concepts and possible outcomes. Like in the Born-Oppenheimer treatment [5.1], we start with the Hamiltonian which contains the interactions of the electrons, the nuclei and the radiation field:

$$H' = \frac{1}{2m}\sum_{\ell}\left[\frac{\hbar}{i}\nabla_{\ell}^{(e)} - e\vec{A}(g_{\ell})\right]^2 + \sum_{j}\frac{1}{2M_j}\left[\frac{\hbar}{i}\nabla_{j}^{(n)} - Z_j e\vec{A}(Q_j)\right]^2 + v(q,Q) + \hbar\omega a^{+}a \quad . \tag{5.116}$$

Here, $(\hbar/i)\nabla_e^{(e)}$ and $(\hbar/i)\nabla_j^{(n)}$ are the linear momenta of electrons and nuclei, respectively, $\vec{A}$ is the vector potential through which the radiation field is described, q and Q are electronic and nuclear coordinates, e and m are the electronic charge and mass, $Z_j e$ and M_j are the nuclear charge and mass, v(q,Q) is the Coulomb interaction between the various charges and the last term stands for the energy of the free radiation field where a and a^+ are the annihilation and creation operators and ω is the frequency associated with the field. In the following, a single-mode field is assumed.

The above Hamiltonian is inconvenient for practical application, and therefore one performs a *contact transformation* [5.109-111].

If ψ stands for the solution of

$$H'\psi = E\psi \quad , \tag{5.117}$$

then, applying a new wave function χ such that

$$\psi = S\chi \quad , \tag{5.118}$$

it is readily seen that χ is a solution of

$$H\chi = E\chi \tag{5.119}$$

where

$$H = S^{-1}H'S \quad . \tag{5.120}$$

In the following, it is assumed that the wave number k is small compared to the range of the interaction, so that $\vec{A}$ can be written as (the dipole approximation)

$$\vec{A} = i\vec{A}_0(a - a^+) \quad . \tag{5.121}$$

Now, if S is written as a product

$$S = \prod_t S_t \tag{5.122}$$

where t runs over all the charged particles (both electrons and nuclei) in the system, a convenient choice for S_t is

$$S_t = \exp\left[\frac{Z_t e}{\hbar}(\vec{A}_0 \cdot \vec{\rho}_t)(a - a^+)\right] \quad , \tag{5.123}$$

where the product in parentheses is a scalar product and $\bar{\rho}_t$ stands for the coordinate of the tth particle.

Applying this transformation and retaining only first-order terms, H takes the more familiar form

$$H = -\frac{\hbar^2}{2m}\sum_{\ell} \nabla_{\ell}^{(e)^2} - \frac{\hbar^2}{2}\sum_{j}\frac{1}{M_j}\nabla_j^{(n)^2} + V(q,Q) + \hbar\omega a^+ a$$
$$+ e\left[\sum_{\ell}(\vec{E}\cdot\vec{q}_{\ell}) - \sum_{j} Z_j(\vec{E}\cdot\vec{Q}_j)\right] \tag{5.124}$$

where $\vec{E}(\rho)$, the electric component of the field, is given as

$$\vec{E} = \frac{\partial A}{\partial t} = -\frac{i}{\hbar}[H_{rad},\vec{A}] \tag{5.125}$$

or, see (5.121),

$$\vec{E} = \omega\bar{A}(0)(a + a^+) \quad . \tag{5.126}$$

The next step is to employ the usual close-coupling technique. Accordingly, the total wave function is written in the form

$$\psi(q,Q) = \sum_{n} |n> \underset{\sim}{\zeta}^* \underset{\sim}{\chi}_n \quad , \tag{5.127}$$

where $|n>$ is the nth eigenstate of the radiation field associated with the eigenvalue $n\hbar\omega$, $\underset{\sim}{\zeta}^*$ is the electronic wave function vector which may (adiabatic) or may not (diabatic) depend on the nuclear coordinates, and χ_{ni} is the nuclear wave function associated with both the ith electronic state and the nth state of the radiation field. (Since the only effect of the field is to shift each state by $n\hbar\omega$, the electronic eigenfunctions are not affected and consequently do not depend on n).

Substitution of (5.127) in (5.124) and integration over the electronic and radiation field coordinates leads to the coupled equations for the nuclear part:

$$T_N\underset{\sim}{\chi}_n + (n\hbar\omega - E)\underset{\sim}{\chi}_n + \underset{\approx}{U}\underset{\sim}{\chi}_n + (\vec{E}\cdot\vec{\underset{\approx}{\mu}})(\sqrt{n+1}\underset{\sim}{\chi}_{n+1} + \sqrt{n}\underset{\sim}{\chi}_{n-1}) = 0 \tag{5.128}$$

where for simplicity, the diabatic representation is employed. Here, $\underset{\approx}{\mu}$ is the dipole moment matrix defined as

$$\mu_{ij} = e\sum_{\ell} <\zeta_i|\vec{q}_{\ell}|\zeta_j> \quad . \tag{5.129}$$

T_n is the nuclear kinetic energy operator and $\underset{\approx}{U}$ is the diabatic potential matrix. In all the applications so far, two "photonic" states were considered. These studies indicated once again that resonance transitions are favorable. Thus, if a certain vibrational level in a given electronic state is shifted (by an amount $\hbar\omega$) in such

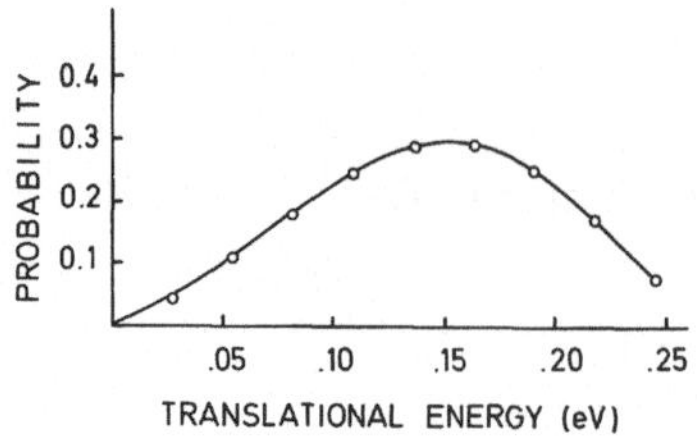

Fig.5.16. The electronic inelastic transition probability of the reaction

$$B_r(^2P_{3/2})+H_2(v_i=0)+h\omega \rightarrow Br(^2P_{1/2})+H_2(v_f=0)$$

as a function of translational energy

a way as to come energetically close to a vibrational state in the other electronic state, this particular (v, e, n) → (v', e', n-1) transition is enhanced considerably. Figure 5.16 shows results for the reaction

$$Br(^2P_{3/2}) + H_2(v_i = 0) + \hbar\omega \rightarrow Br(^2P_{1/2}) + H_2(v_f = 0) \quad . \tag{5.130}$$

As can be seen from Fig.5.11, this process is very unlikely without the field but becomes considerably enhanced once the field is turned on. The opposite situation might occur too, namely, the field might weaken certain transitions which would otherwise have been much stronger [5.105,112]. Most of the treatments were carried out for nonreactive systems and only one reactive model study has been reported [5.113]. An interesting idea was discussed recently by OREL and MILLER [5.114] who considered the possibility of enhancing the H + H_2 reaction by lowering the activation energy due to an interaction between the radiation field and the molecular system. It was argued that such an interaction could occur when the three H atoms were in close proximity and behaved, for a short time, as a (collinear) three-atom molecule which could become infrared active due to the asymmetric stretch. This process is plausible, even though neither the reagents nor the products are infrared active. The possibility of the potential barrier of a reactive system being lowered due to the existence of a (temporary or permanent) upper electronic state was discussed and treated in [5.64,65] and could apply to this and other similar studies [5.113].

5.5 Summary and Conclusions

This review has described what has been done so far on electronic transitions taking place during an atom-diatom collision, applying quantum mechanical techniques. The field is relatively new - the first extensive numerical treatments appeared around 1975 - and this explains why so little has been achieved. As an example, I refer to the diabatic versus the adiabatic approach. The two frameworks have their advantages and disadvantages, and these were considered to a certain extent for collinear systems. However, as far as three-dimensional systems are concerned, only little is known. The quenching process of an excited fluorine by H_2 was studied by two groups - one study was performed within the adiabatic framework and the other within the diabatic. The results differ by one order of magnitude. The question to be raised is whether this large difference is due to the different potential surfaces applied by the two groups or to some confusion regarding the two frameworks. Therefore, one important subject in future studies should relate to a deeper understanding of these representations in three dimensions and of their interrelations. Unless such studies are performed, approximations done by ignoring various terms cannot be estimated and consequently the relevance of the final numerical outcomes also becomes unclear.

Another situation lacking clarity exists regarding the classical and the semiclassical approaches applied to electronic transitions between surfaces. Two such methods are available; one of them applies the Landau-Zener formula and the other the Stückelberg method. The results obtained with these methods were compared with exact quantum mechanical results on only very rare occasions and, although the agreement was not encouraging, still no further attempts were made to study this subject.

The main reason for the late start in the study on electronic transitions and the slow progress made is connected with the complexity of electronic transitions. The treatment of electronic transitions, more than any other process, calls for the development of approximate methods. I consider that such processes for nonreactive collisions could be dealt with reliably using the distorted wave Born approximation. In the few cases where this method was applied and the outcome was examined with respect to the results of the exact treatment, the comparison was rather encouraging. As for electronic transitions in reactive collisions, they seem very likely to be treated quite reliably within the Infinite Order Sudden Approximation which has recently been shown to be adequate in calculations of reactive (single surface) cross sections.

Acknowledgments

I want to thank Profs. H. Kruger, F. Linder and P. McGuire for inviting me to stay at the Department of Physics of the University of Kaiserslautern (Fed. Rep. of Germany) for the period from July to December 1980, during which this composition was completed. I would also like to thank the "Deutsche Forschungsgemeinschaft" which, within the framework of the Sonderforschungsbereich 91 "Energietransfer bei Atomaren und Molekularen Stossprozessen", partially supported this work. And finally, I want to thank my wife Malvine for editing this review and helping me to prepare the manuscript.

References

5.1 M. Born, J.R. Oppenheimer: Ann. Phys. (Leipzig) *84*, 457 (1927)
5.2 C. Zener: Proc. Roy. Soc. A*137*, 696 (1932)
5.3 L.D. Landau: Phys. Z. Soviet Union *2*, 46 (1932)
5.4 E.C.G. Stückelberg: Helv. Phys. Acta *5*, 369 (1932)
5.5 W. Lichten: Phys. Rev. *131*, 229 (1963)

5.6 E.E. Nikitin: In *Chemische Elementar Prozesse,* ed. by H. Hartman (Springer, Berlin, Heidelberg, New York 1968)
5.7 N. Rosen, C. Zener: Phys. Rev. *40*, 502 (1932)
5.8 F.T. Smith: Phys. Rev. *179*, 111 (1969)
5.9 M.S. Child: Molec. Phys. *20*, 171 (1970)
5.10 Yu.N. Demkov: Sov. Phys. JETP *18*, 138 (1964)
5.11 R. de L. Kronig: *Band Spectra and Molecular Structure* (Cambridge University Press, New York 1930) pp.6-16
5.12 D.R. Bates: Proc. Roy. Soc. A*240*, 437 (1957); A*243*, 15 (1958)
5.13 D.R. Bates: Proc. Roy. Soc. A*257*, 22 (1960)
5.14 W.R. Thorson: J. Chem. Phys. *34*, 1744 (1961); *42*, 3878 (1965)
5.15 D.J. Kouri, C.F. Curtiss: J. Chem. Phys. *44*, 2120 (1966)
5.16 R.T. Pack, J.O. Hirschfelder: J. Chem. Phys. *49*, 4009 (1968); *52*, 521 (1970)
5.17 C. Gaussorgues, C. Le Sech, F. Mosnow-Seeuws, R. McCarroll, A. Riera: J. Phys. B, At. Mol. Phys. *8*, 239 (1975)
5.18 C. Gaussorgues, C. Le Sech, F. Mosnow-Seeuws, R. McCarroll, A. Riera: J. Phys. B. At. Mol. Phys. *8*, 253 (1975)
5.19 G. Gioumousis, D.P. Stevenson: J. Chem. Phys. *29*, 294 (1958)
5.20 F.S. Klein, L. Friedman: J. Chem. Phys. *41*, 1789 (1964)
5.21 A. Henglein, K. Lackmann, B. Knoll: J. Chem. Phys. *43*, 1048 (1965)
5.22 E.A. Gislason: J. Chem. Phys. *57*, 3396 (1972)
5.23 E. Bauer, E.R. Fischer, F.R. Gilmore: J. Chem. Phys. *51*, 4173 (1969)
5.24 E.E. Nikitin, S.Ya. Umanski: Dis. Farad. Soc. Chem. *53*, 7 (1972)
5.25 M.S. Child: Farad. Dis. Chem. Soc. *55*, 30 (1973)
5.26 M.S. Child: *Molecular Collision Theory* (Academic Press, London 1974) pp.161-179
5.27 A. Bjerre, E.E. Nikitin: Chem. Phys. Lett. *1*, 179 (1967)
5.28 J.C. Tully, R.K. Preston: J. Chem. Phys. *55*, 562 (1971)
5.29 J.R. Krenos, R.K. Preston, R. Wolfgang, J.C. Tully: J. Chem. Phys. *60*, 1634 (1974)
5.30 G. Ocs, E. Teloy: J. Chem. Phys. *61*, 4930 (1974)
5.31 W.H. Miller, T.F. George: J. Chem. Phys. *56*, 5637 (1972)
5.32 Y.-W. Lin, T.F. George, M. Morokuma: J. Chem. Phys. *60*, 4311 (1974)
5.33 J.R. Laing, T.F. George, I.H. Zimmermann, Y.-W. Lin: J. Chem. Phys. *63*, 842 (1975)
5.34 D.A. Micha: Adv. Chem. Phys. *30*, 7 (1975)
5.35 J.C. Tully: In *Dynamics of Molecular Collisions, Part B,* ed. by W.H. Miller (Plenum Press, New York 1976) Chap.5
5.36 M.S. Child: In *Atom-Molecule Collision Theory,* ed. by R.B. Bernstein (Plenum Press, New York 1979) Chap .13
5.37 T.F. O'Malley: Adv. At. Mol. Phys. *7*, 223 (1971)
5.38 R.G. Gordon: J. Chem. Phys. *51*, 14 (1969)
5.39 W.N. Sams, D.J. Kouri: J. Chem. Phys. *51*, 4815 (1969)
5.40 M. Baer: Chem. Phys. Lett. *35*, 112 (1975)
5.41 M. Baer: Chem. Phys. *15*, 49 (1976)
5.42 Z.H. Top, M. Baer: J. Chem. Phys. *66*, 1363 (1977)
5.43 Z.H. Top, M. Baer: Chem. Phys. *25*, 1 (1977)
5.44 M. Baer, J.A. Beswick: Phys. Rev. A*19*, 1559 (1979)
5.45 M. Baer, G. Drolshagen, J.P. Toennies: J. Chem. Phys. *73*, 1690 (1980)
5.46 F.O. Ellison: J. Am. Chem. Soc. *85*, 3540, 3544 (1963)
5.47 F.O. Ellison, J.C. Patel: J. Am. Chem. Soc. *86*, 2155 (1964)
5.48 R.K. Preston, J.C. Tully: J. Chem. Phys. *54*, 4297 (1971)
5.49 P.J. Kuntz, A.C. Roach: J. Chem. Soc. Farad. Trans. *II68*, 259 (1972)
5.50 J.C. Tully: J. Chem. Phys. *58*, 1396 (1973)
5.51 Y. Zeiri, M. Shapiro: J. Chem. Phys. *70*, 5264 (1979)
5.52 I. Last, M. Baer: J. Chem. Phys. *75*, 288 (1980)
5.53 I. Last, M. Baer: Chem. Phys. Lett. *73*, 514 (1980)
5.54 I. Last: Chem. Phys. *55*, 237 (1981)
5.55 D.J. Vezzetti, S.I. Rubinow: Ann. Phys. *35*, 373 (1955)
5.56 W.H. Miller: J. Chem. Phys. *43*, 2373 (1968)
5.57 P.L. DeVries, T.F. George: J. Chem. Phys. *67*, 1293 (1968)
5.58 D.L. Miller, R.E. Wyatt: J. Chem. Phys. *67*, 1302 (1977)

5.59 E.E. Nikitin: *Theory of Elementary Atomic and Molecular Processes in Gases* (Clarendon Press, Oxford 1974) p.148
5.60 F. Rebentrost, W.A. Lester: J. Chem. Phys. *67*, 1977 (1977)
5.61 Y. Haas, R.D. Levine, G. Stein: Chem. Phys. Lett. *15*, 7 (1972)
5.62 G.L. Hofacker, R.D. Levine: Chem. Phys. Lett. *9*, 617 (1971)
5.63 H. Nakamura: Mol. Phys. *26*, 673 (1973)
5.64 Z.H. Top, M. Baer: Chem. Phys. *10*, 95 (1975)
5.65 Z.H. Top, M. Baer: Chem. Phys. *16*, 447 (1976)
5.66 M. Baer, J.A. Beswick: Chem. Phys. Lett. *51*, 360 (1977)
5.67 P.J. Kuntz, A.C. Roach: J. Chem. Soc. Farad. Trans. *II68*, 259 (1972)
5.68 R.M. Billota, F.N. Preuminger, J.M. Farrar: Chem. Phys. Lett. *74*, 95 (1980)
5.69 R.M. Billota, F.N. Preuminger, J.M. Farrar: J. Chem. Phys. *73*, 1637 (1980)
5.70 J.M. Bowman, S.C. Leasure, A. Kuppermann: Chem. Phys. Lett. *43*, 374 (1976)
5.71 C.D. Jonah, R.N. Zare, Ch. Ottinger: J. Chem. Phys. *56*, 263 (1972)
5.72 D.J. Wren, M. Menziger: J. Chem. Phys. *63*, 4557 (1975)
5.73 I.H. Zimmerman, M. Baer, T.F. George: J. Chem. Phys. *71*, 4132 (1979)
5.74 I.H. Zimmerman, T.F. George: Chem. Phys. *7*, 323 (1975)
5.75 D.F. Feng, E.R. Grant, J.W. Root: J. Chem. Phys. *64*, 3450 (1976)
5.76 N.C. Blais, D.G. Truhlar: J. Chem. Phys. *69*, 846 (1978)
5.77 A. Komornicki, K. Morokuma, T.F. George: J. Chem. Phys. *67*, 5012 (1977)
5.78 R.E. Wyatt, R.B. Walker: J. Chem. Phys. *70*, 1501 (1979)
5.79 P.L. DeVries, T.F. George: J. Chem. Phys. *66*, 2421 (1977)
5.80 F. Rebentrost, W.A. Lester: J. Chem. Phys. *63*, 3737 (1975)
5.81 F. Rebentrost, W.A. Lester: J. Chem. Phys. *64*, 3879 (1976)
5.82 S.I. Chu, A. Dalgarno: J. Chem. Phys. *62*, 4009 (1975)
5.83 B. Amaee, C. Botcher: J. Phys. B*11*, 1249 (1978)
5.84 P. McGuire, J.C. Bellum: J. Chem. Phys. *71*, 1975 (1979)
5.85 P. Botschwina, W. Mayer: Unpublished
5.86 I.V. Hertel, H. Hofmann, K.A. Rost: Chem. Phys. Lett. *46*, 163 (1977)
5.87 H. Pelzer, E.P. Wigner: Z. Phys. Chem. B*15*, 445 (1932)
5.88 H. Eyring, M. Polanyi: Z. Phys. Chem. B*12*, 279 (1931)
5.89 C. Eckart: Phys. Rev. *35*, 1303 (1930)
5.90 E.P. Wigner: Z. Phys. Chem. B*19*, 203 (1933)
5.91 R.P. Bell: Proc. Roy. Soc. A*139*, 466 (1933)
5.92 M.S. Child: Mol. Phys. *16*, 313 (1969)
5.93 M.S. Child: Mol. Phys. *28*, 495 (1974)
5.94 E.E. Nikitin: Adv. Quantum Chem. *5*, 135 (1970)
5.95 D.S.F. Crothers: J. Phys. B*6*, 1418 (1973)
5.96 M. Baer, M.S. Child: Mol. Phys. *36*, 1449 (1978)
5.97 M.S. Child: *Molecular Collision Theory* (Academic Press, London 1974) p.122
5.98 S.C. Miller, R.H. Good: Phys. Rev. *91*, 174 (1954)
5.99 M.S. Child: Mol. Phys. *29*, 1421 (1975)
5.100 R.D. Levine, G.G. Balint-Kurti: Chem. Phys. Lett. *6*, 101 (1970)
5.101 R.D. Levine: Mol. Phys. *22*, 497 (1971)
5.102 J.R. Laing, J.M. Yuan, I.H. Zimmerman, P.L. DeVries, T.F. George: J. Chem. Phys. *66*, 2801 (1977)
5.103 H.J. Korsch, H. Kruger: Mol. Phys. *39*, 51 (1980)
5.104 M.S. Child, M. Baer: J. Chem. Phys. *74*, 2832 (1981)
5.105 I.H. Zimmerman, J.M. Yuan, T.F. George: J. Chem. Phys. *66*, 2638 (1977)
5.106 N.M. Kroll, K.M. Watson: Phys. Rev. A*8*, 804 (1973); A*13*, 1018 (1976)
5.107 A.M.F. Lau: Phys. Rev. A*13*, 139 (1976); A*14*, 279 (1976)
5.108 J.I. Gersten, M.H. Mittleman: Phys. Rev. A*12*, 1840 (1975)
5.109 M. Goppert Mayer: Ann. Phys. Leipzig *9*, 273 (1931)
5.110 P.I. Richards: Phys. Rev. *73*, 254 (1949)
5.111 E.A. Power, S. Zienau: Phil. Trans. Roy. Soc. *251*A, 427 (1959)
5.112 T.F. George, I.H. Zimmerman, P.L. DeVries, J.M. Yuan, Kai-Shae Lam, J.C. Bellum, H.W. Lett, M.S. Slatsky, J.T. Lin: In *Chemical and Biochemical Applications of Lasers*, Vol.4, ed. by C.B. Moore (Academic Press, New York 1979), in press
5.113 J.C. Light, A. Altenberger-Siczek: J. Chem. Phys. *70*, 4108 (1979)
5.114 A.E. Orel, W.H. Miller: Chem. Phys. Lett *57*, 362 (1978)

Subject Index